Fun Packaging

Fun Packaging
by Miquel Abellán

English edition published in 2014 by Gingko Press Inc.
1321 Fifth Street, Berkeley, CA 94710, USA
www.gingkopress.com

First published in Japan in 2014 by
Graphic-sha Publishing Co., Ltd.
1-14-17 Kudankita, Chiyoda-ku,
Tokyo 102-0073, Japan

Creative staff
Author: Miquel Abellán (Arts Visuals)
Original idea and editing: Louis Bou (Mista Studio)
Design and layout: Miquel Abellán (Arts Visuals)
Production: Kumiko Sakamoto (Graphic-sha Publishing)

Printed and bound in China

ISBN 978-1-58423-539-2

Fun Packaging

GINGKO PRESS

index

index

Even though the goal of modern retail packaging design is to encourage potential buyers to purchase the product, funny packages can also be a great source of inspiration. Fun Packaging is a comprehensive compilation of international packaging projects, where visual graphics and creativity bring us into a world where fun is the main player in the project itself; improving and easing communication, but with a smart sense of humor.
This humorous touch can communicate the same goals – no doubt about it – as more common or high-end packaging designs, providing a consumer experience that goes beyond the descriptive and conceptual by engaging the buyer in an amusing experience.

Have fun! Get inspired! Smile! We're talking about packaging design! :-D

AU YEAH!
AMERICAN
PALE ALE
CERVEZA
ARTESANAL
4.5%

Packages for bottles and cans

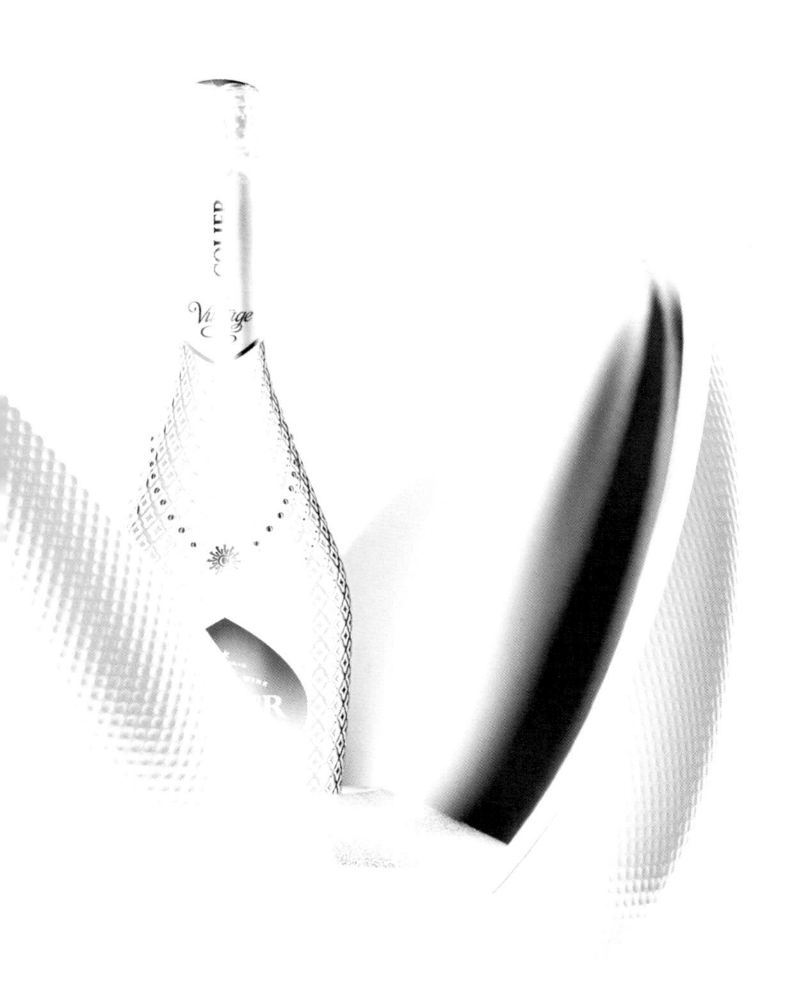

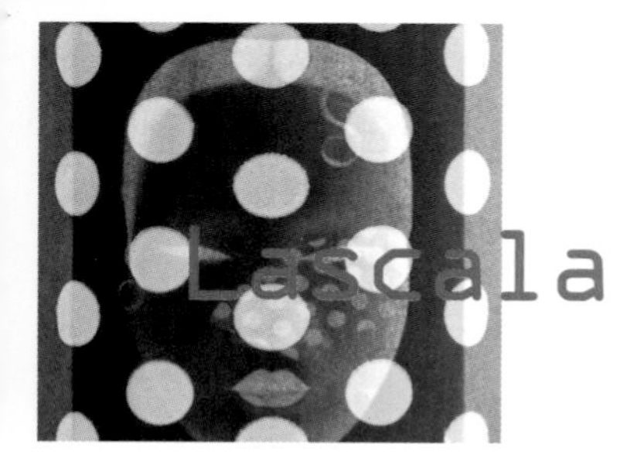

Lascala

studio **Eduardo del Fraile**
Murcia Spain
designer Eduardo d Fraile
www.eduardodfraile.com

A long-standing Spanish winemaker wanted to enter the Chinese market. This process had to be as harmonic as possible. French wines, which are the most popular in China, usually choose to keep their identity 100%, showing their authenticity. In this case, the concept was the merging of Western and Eastern cultures. The Chinese call the theater LASCALA, so a painted face imitating a theater mask with Eastern eyes was used to represent China and the merger is represented by LA PEINETA (the ornamental comb) Ros Wine, EL ABANICO (the hand fan) White Wine and LA BAILAORA DE FLAMENCO (the Flamenco dancer) Red Wine.
Products that are not originally manufactured in China cannot bear Chinese writing on the front of the packaging. Our solution was to use typography in a vertical axis, mimicing the Chinese alphabet.

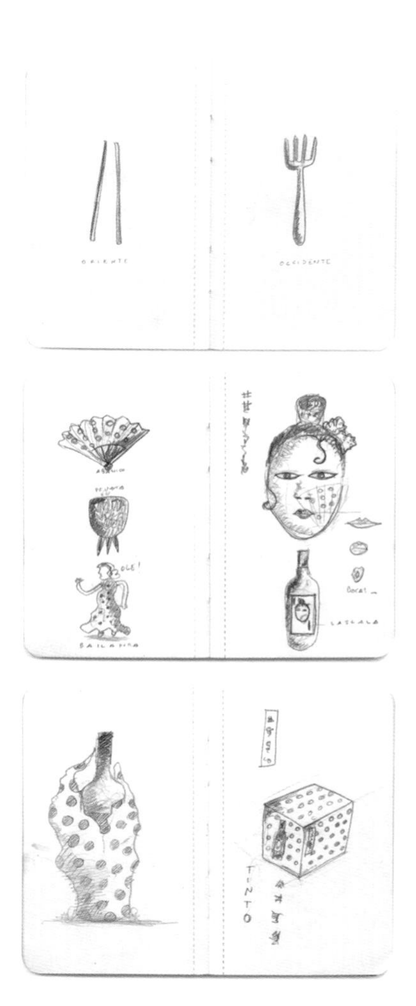

TINTO
LASCALA
ROSADO
LASCALA
BLANCO
LASCALA

LASCALA
LASCALA

LASCALA

BLANCO

Pistonhead Crude Oil

studio **Neumeister Strategic Design AB**
Stockholm Sweden
creative director Henrik Hallberg
designers Lachlan Bullock, Per Torell, Gustav Schultz
www.neumeister.se

Pistonhead is a brand from Brutal Brewing that has previously launched a number of conceptual beer variants - all inspired by Kustom Kulture aesthetics. Neumeister was assigned to create an expression in line with the current identity while allowing growth for future brand extensions. There was also a need to stretch the brand to create a different kind of alcoholic beverage.

Pistonhead Crude Oil is a chili based liquorice shot. Its iconic Calavera skull is the main sender and works as an integrated sub-brand to the Pistonhead name. Printing technique, typography and handling of details were more essential than for any prior line extension. The Kustom Kulture aesthetics simply had to be enhanced even further, preparing the brand to enter a wider spectrum of product categories - without losing its recognition.

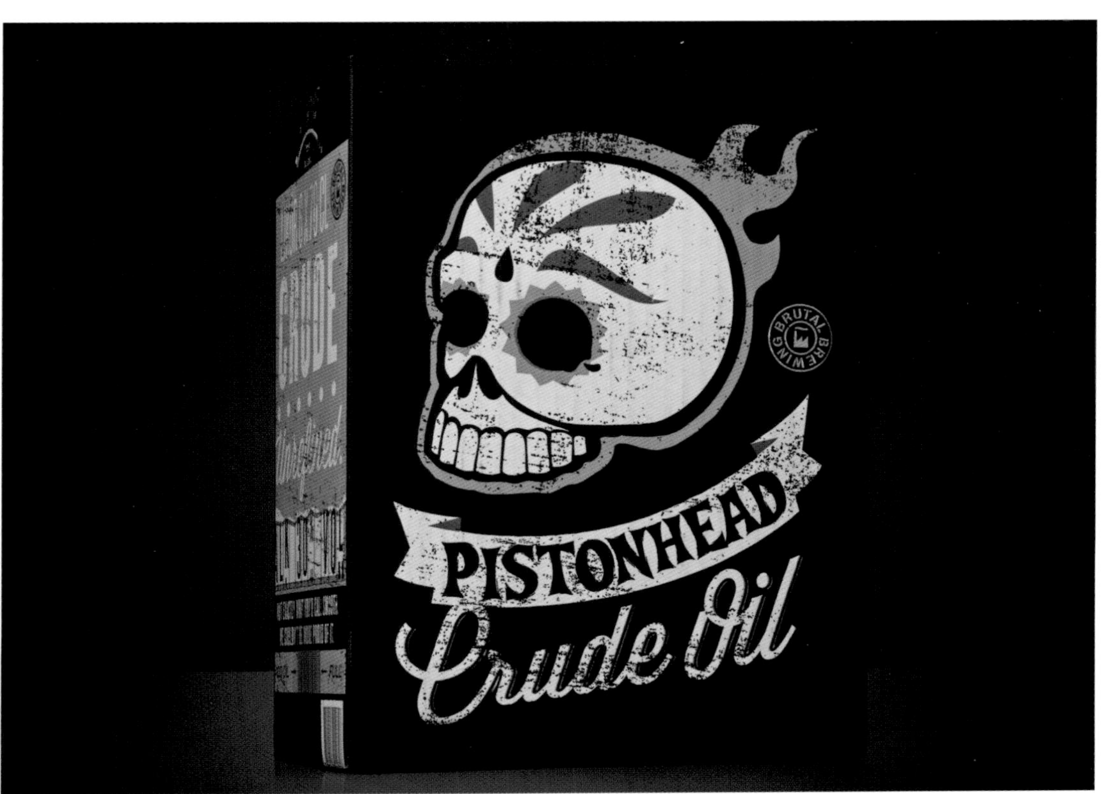
BRUTAL BREWING
PISTONHEAD
Crude Oil
CRUDE
Unrefined.

PISTONHEAD
Crude Oil
LIKÖR
70 CL
CRUDE
Unrefined.
20 40 60
OIL LEVEL
NOT EXACTLY WHAT YOU'D CALL SMOOTH
WE COULDN'T BE MORE PROUD OF IT
ALK 30% VOL
I TRAFIKEN?
AVSTÅ ALKOHOL!

The Roller Derby Collection

agency **Studio Lost & Found**
Perth Australia
designer & illustrator Daniel McKeating
photographer Daniel McKeating
www.studiolostandfound.com

Hand-made in Margaret River by Tash Arthur, The Roller Derby Collection includes a red fortified wine called Ruby Slipper and a white fortified wine called Glass Slipper. The labels feature illustrations of Dorothy Gale and Cinderella, but with a unique Roller Derby twist.
Both products are packaged in premium French glass with authentic Portuguese stopper corks. The labels were printed in four-colour process on Fasson Estate 8 premium uncoated paper, with a clear high-build silkscreen varnish.

THE
ROLLER DERBY
COLLECTION
HAND MADE IN MARGARET RIVER BY TASH ARTHUR

RED
FORTIFIED
WINE
RUBY
SLIPPER

WHITE
FORTIFIED
WINE
GLASS
SLIPPER

agency **Reynolds & Reyner**
Kiev Ukraine
creative directors Artyom Kulik,
Alexander Andreyev
www.reynoldsandreyner.com

Colier targets business women as worldwide consumers of sparkling wines. The series is limited and 5 sets of premium Vintage Brut come in handmade bottles packaged within a cocoon container. The cocoon's weight is concentrated at the bottom of the package to lend it stability. The bottle comes with a necklace around its neck, evoking the most expensive and elegant of women's jewelry. The concept is to reinforce a sophisticated and highly artistic form, emphasizing femininity and a refined style. As a bonus, the container maintains the temperature so that the champagne stays cold for a long time.

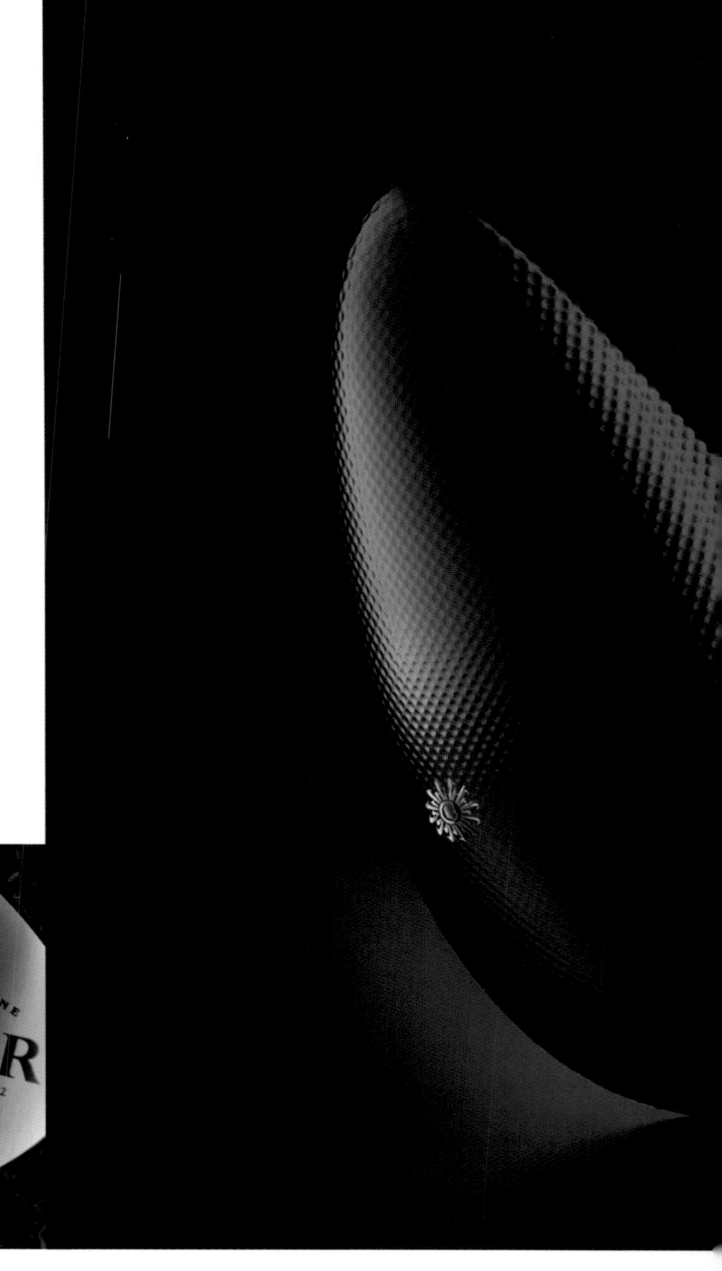

Plagios

agency **Beetroot Design Group**
Thessaloniki Greece
designers Vagelis Liakos, Alexis Nikou & Yiannis Charalambopoulos
www.beetroot.gr

The design for the labels of "Plagios" wines ("Plagios, white" and "Plagios, red"), produced by the Biblia Chora Winery, borrowed conceptual inspiration from the name by evoking the motion (plagios means sideways in Greek) of the bishop piece on a chessboard. The two colors of the chessboard were used on the label, and positioned at an angle, so that the painted inner white squares are offset by the bottle's dark glass. The colors are echoed in the bottle's seal to indicate the type of wine: a white seal for white wine and black for red. A delicate golden symbol in the shape of a bishop-piece, the winery's crest and an inscription of the wine's name, complete the elegant composition of the label.

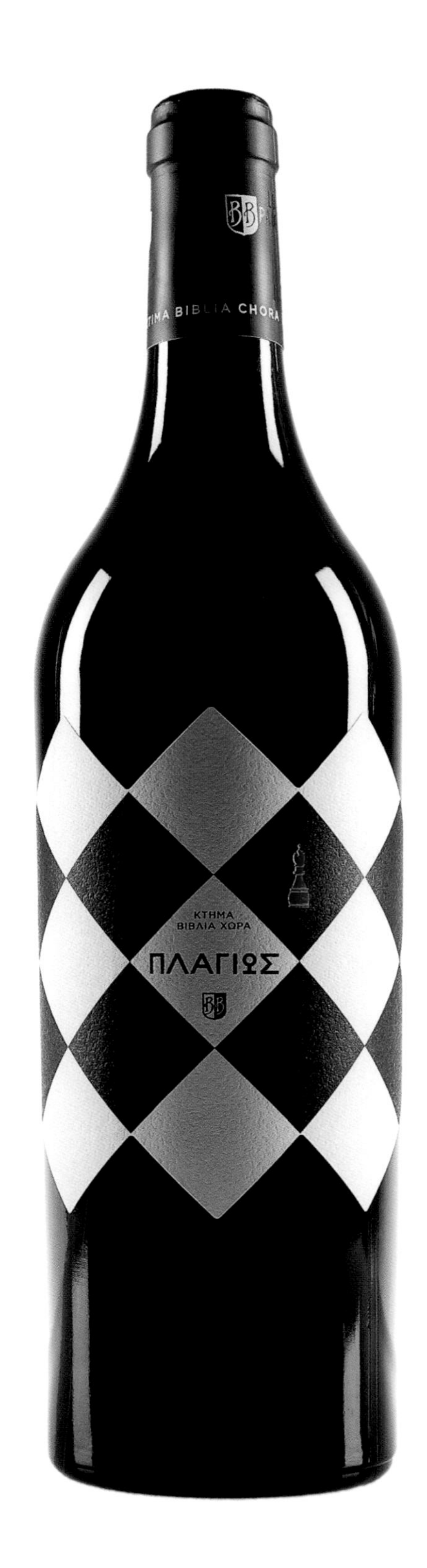

agency **Studio Lost & Found**
Perth Australia
designer & illustrator Daniel McKeating
www.studiolostandfound.com

Knee Deep is an award-winning winery and restaurant based in Wilyabrup, Margaret River, Western Australia. Project scope included the refinement of their existing entry-level, mid-tier, and limited release range packaging.

Au yeah!

studio **mLlongo**
Valencia Spain
art director Marisa Llongo
design / illustration Studio mLlongo
www.mllongo.com

Au yeah, is a Valencian beer with an American style, dixie inspiration, Texas hops and authentic flavor. The mixture of cultures is presented with a touch of humor in the naming, a unique graphic design and handmade typography.

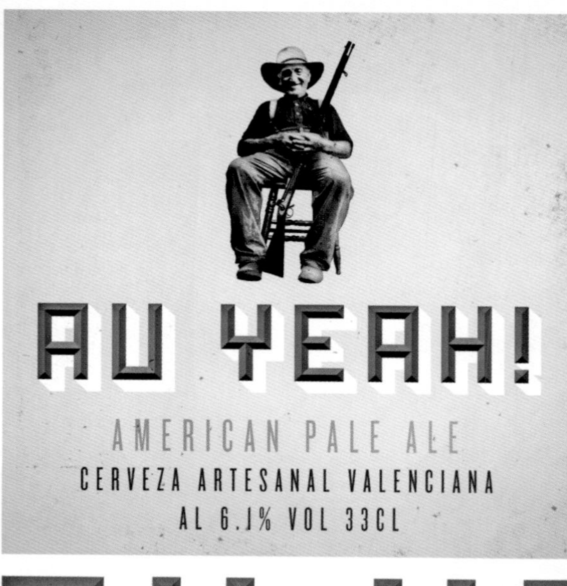

AU YEAH!

AU YEAH!
AMERICAN
PALE ALE
CERVEZA
ARTESANAL
VALENCIANA
4.5%
VOL 33CL

Ese espíritu mueve esta cerveza
Paradigmas los justos
Hemos explorado el panorama
cervecero americano para
conseguir una cerveza de alta
fermentación de carácter
yankee. Color dorado, muy fresca,
aromática y diferente.
Ingredientes Malta Pale Ale
lúpulo americano y levaduras
seleccionadas. Sólo ingredientes naturales
para conseguir un sabor limpio y afrutado
Fermentada por segunda vez en
botella buscando el punto justo de
maduración y presión. Sin pasteurizar
ni filtrar. 100% natural.
Elaborada por IN EXTREMIS S.L.
P.I. Oliveral - Ribarroja del Turia
Reg Sanitario 30.11335/V
8 437011 135544

CUBEN Space
Johannesburg South Africa
London UK
designers Marcel Buerkle / Simon C. Page
www.behance.net/marcel_b
www.behance.net/simoncpage

This concept packaging for fruit wine is inspired by Simon C. Page's wonderful CUBEN project. The boxes and labels feature pattern designs by both Simon C. Page and Marcel Buerkle.

LUX FRUCTUS
APPLE

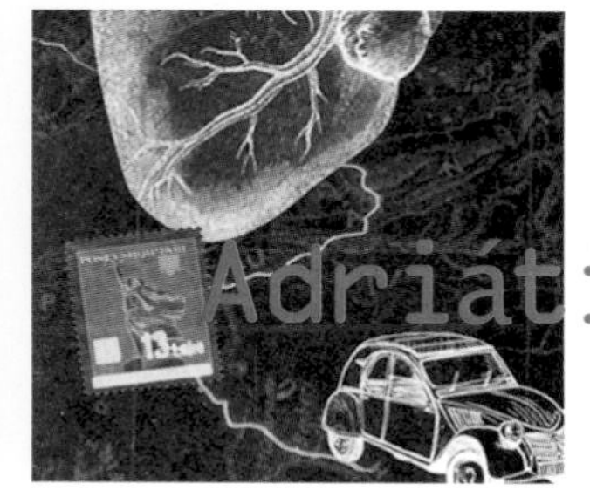

Adriático

studio **Julia Castaño**
Barcelona Spain
designers Julia Castaño, Marta González, Yesika Aguín
www.juliacastano.com
dropr.com/martagpalacios
www.domestika.org/portfolios/yesikina

Adriatico's packaging is designed to send the consumer on a virtual cycling trip around the Adriatic coast. Each of the trip's stages is portrayed by different labels: the beginning shows the anticipation of the trip; the middle shows the anecdotes and experiences that happened along the way; and the end of the trip is about saying goodbye to the adventure and gathering memories from the trip.
The three wine bottles are united through a hand-drawn line and each label adds a new bicycle seat, which symbolizes the people you have met on the way.

ADRIÁTICO
2003

ADRIÁTICO
2003

ADRIÁTICO
2008

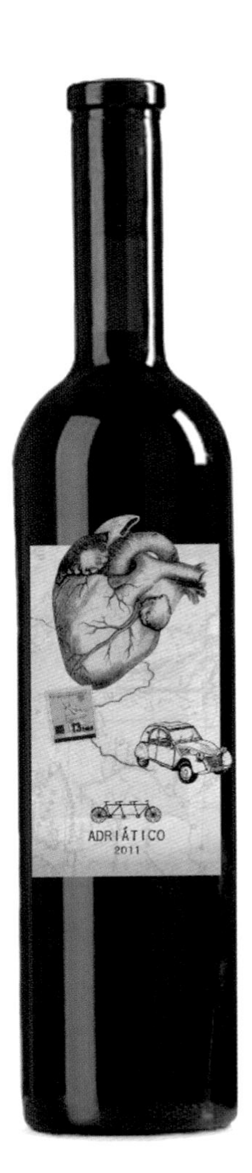
ADRIÁTICO
2011

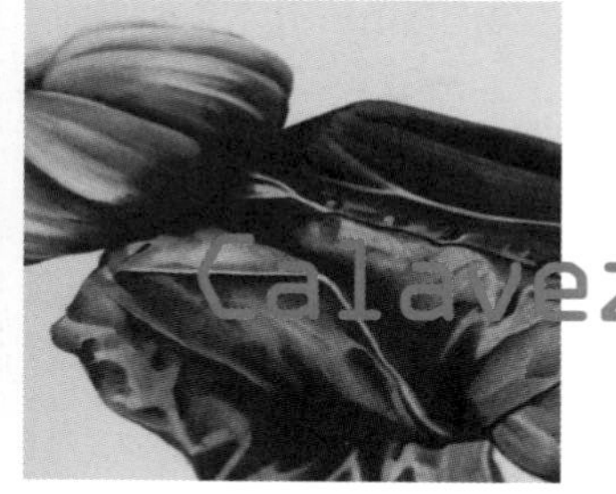

Calaveza

studio **mLlongo**
Valencia Spain
art director Marisa Llongo
design / illustration Studio mLlongo
www.mllongo.com

Calaveza is a special beer that gets its name from the mixture of the Spanish words 'pumpkin' (calabaza) and 'beer' (cerveza). Several illustrations have been developed for the packaging, as if for a family portrait, but the heads have been replaced by pumpkins resulting in a visual game that links the identity of the beer with its principal ingredient. Several labels have been used in the packaging, each one with different color, material and texture. All together, they make up the meaning of the beer.

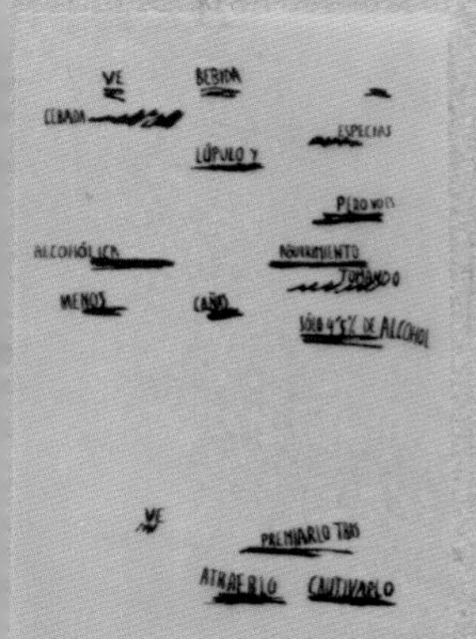

calabaza *f* **1** Fruta alimenticia de la cucurbitácea, compuesta en un 90% de agua, conteniendo también pepitas y, por supuesto, pulpa de calabaza. Tiene un sabor dulce y un color anaranjado. Es diurética, saludable y nutritiva y es muy aconsejable su ingesta en ayunas. Es antipirética, previene el estreñimiento y calma el dolor de cabeza (infusionando más de tres tazas). Posee un alto contenido en vitaminas y poca grasa por lo que su consumo diario es muy recomendable. Gastronómicamente se suele emplear como acompañante en las comidas aunque sola resulta exquisita. Se encuentra disponible en otoño y es protagonista en la ornamentación de fiestas como Navidad y Halloween.

dar calabazas a alguien.

1. *loc. verb. coloq.* Reprobarlo en un examen.

2. *loc. verb. coloq.* Desairarlo o rechazarlo cuando requiere de amores.

CALAVEZA

deli_rant.

diseñada por estudio mllongo

Spanish orange juice

studio **P&W Design Consultants**
London UK
designer Iain Dobson
creative directors Simon Pemberton & Adrian Whitefoord
www.p-and-w.com

The design of Tesco's orange juice packaging reflects the origin of the fruit whilst keeping it light hearted and distinctive from the array of other orange juices on the shelves. The image of the Matador was used to create an iconic symbol - integrating the photographic orange within the illustrations for a sense of fun, while helping the consumer quickly differentiate between the juice varieties coming with pulp (bits) and without (smooth).

Best before
New!
TESCO
SPANISH
ORANGE
100% Pure squeezed juice
Not from concentrate
Smooth
1 of
5 a day
250ml contains
Calories
120
6%
Sugar
26.5g
29%
Fat
trace
<1%
Saturates
0g
0%
Salt
trace
<1%
of your guideline daily amount

Best before
New!
TESCO
SPANISH
ORANGE
100% Pure squeezed juice
Not from concentrate
With Bits
1 of
5 a day
250ml contains
Calories
120
6%
Sugar
26.5g
29%
Fat
trace
<1%
Saturates
0g
0%
Salt
trace
<1%
of your guideline daily amount

agency **Studio Lost & Found**
Perth Australia
designer & illustrator Daniel McKeating
photographer Jay Heifetz
www.studiolostandfound.com

The Quest is a premium range of wines by renowned Margaret River producer, Chalice Bridge Estate. The concept draws inspiration from the tales of the Knights Templar and their relentless quest for the Holy Grail. The Quest also pays homage to Chalice Bridge's limited release tier, The Chalice.
The Quest wine labels were printed on Fasson Antarctic White premium vellum paper in 6-color, offset with gold foil and a clear silkscreen varnish.

CHALICE BRIDGE
— THE —
QUEST
Single Vineyard
MARGARET RIVER
2008 MERLOT

CHALICE BRIDGE
— THE —
QUEST
Single Vineyard
MARGARET RIVER
2008 CABERNET SAUVIGNON

CHALICE BRIDGE
— THE —
QUEST
Single Vineyard
MARGARET RIVER
2008 SHIRAZ

Left Hand Brewing

agency **Moxie Sozo**
Boulder, CO USA
designers Nate Dyer, Charles Bloom
www.moxiesozo.com

Longmont, Colorado craft brewery Left Hand had the goods, but was in need of a dynamite website and packaging to match the bold and adventurous brews the company had to offer. It was no surprise that Moxie Sozo was so eager to work with a brand that had such, well, moxie! Founder Dick Doore's homebrew kit spawned the humble beginnings of the brewery and Left Hand quickly became a local favorite. As the brewery began to expand its reach across the nation, Moxie Sozo redesigned their packaging and website, creating a very distinct look and feel for the exceptional beer.
As a result, the brewer promptly ran out of beer within eight months of the labels' debut, forcing Left Hand to increase production to accommodate demand. Could there be a better problem to have? We think not.

LEFT HAND BREWING CO.
LEFT HAND BREWING CO. PRESENTS
WARRIOR
IPA
BREWED ONLY ONCE A YEAR
WITH COLORADO FRESH HOPS
INDIA PALE ALE

LEFT HAND BREWING CO. PRESENTS
WARRIOR
IPA
BREWED ONLY ONCE A YEAR
WITH COLORADO FRESH HOPS
INDIA PALE ALE

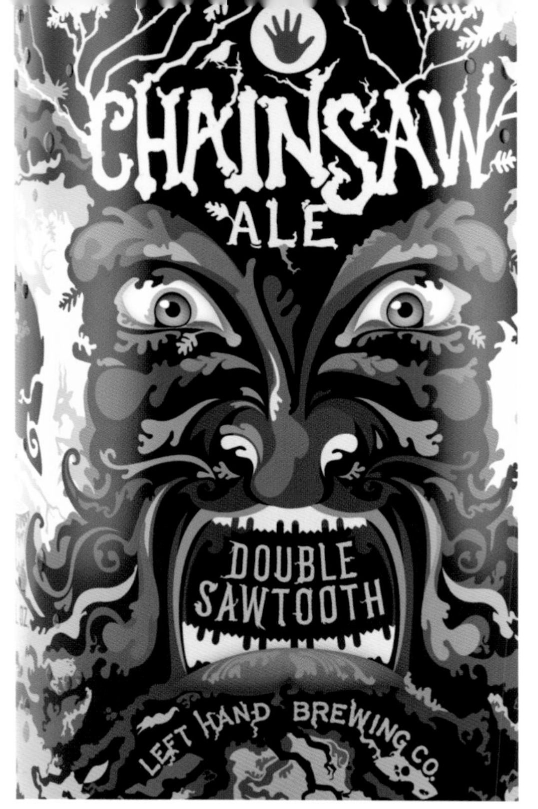
CHAINSAW
ALE
DOUBLE
SAWTOOTH
LEFT HAND BREWING CO.

LEFT HAND BREWING CO.
GOOD
JUJU
BEER BREWED WITH
GINGER
A REFRESHING
FRIVOLITY

LEFT HAND BREWING CO.
LEFT HAND BREWING CO. PRESENTS
WARRIOR
IPA
WITH COLORADO FRESH HOPS
INDIA PALE ALE
TWIN
SISTERS
DOUBLE
IPA
INDIA
PALE ALE
LEFT HAND
BREWING CO.
400
POUND
MONKEY
ENGLISH-STYLE
INDIA PALE ALE
GOOD
JUJU
BEER BREWED WITH
GINGER
A REFRESHING
FRIVOLITY
ALE
DOUBLE
SAWTOOTH

Milk Stout
12 FL. OZ.
left hand brewing co.
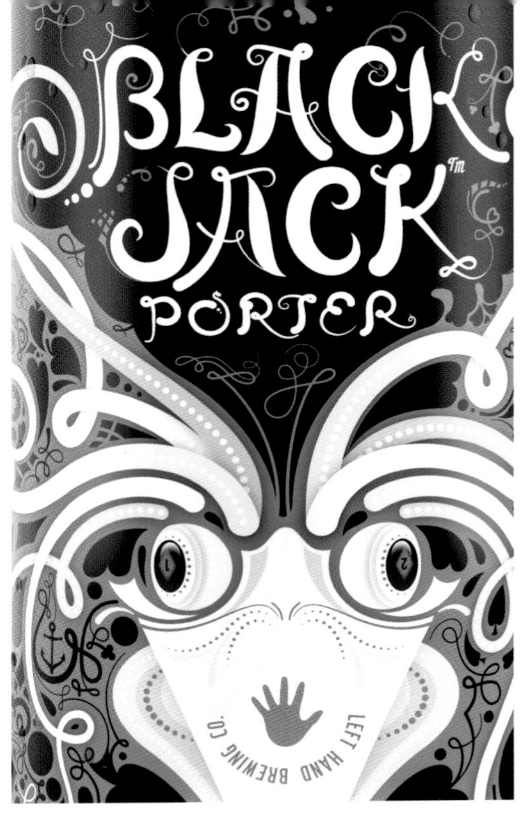
Black Jack
Porter
Left Hand Brewing Co

Sawtooth Ale
Left Hand Brewing Co.
Black Jack Porter
Oktoberfest
Märzen Lager
Milk Stout
left hand brewing co.
fade to black
Ale
12 fluid ounces

Left Hand Brewing

agency **Moxie Sozo**
Boulder, CO USA
designers Nate Dyer, Charles Bloom
www.moxiesozo.com

LEFT HAND
BREWING CO.
fade
left hand
to
brewing co.
black
ALE
12
fluid ounces

fade
left hand
to
brewing co.
black
ALE
12

studio **Bayley Design**
London UK
designer William Suckling
www.behance.net/bayleydesign

In British homes, red and brown sauce are a given. Every house has them and everyone uses them. The brief was to design packaging for sauce that would come in a Harvey Nichols food basket. The Victorian engravings of the man and woman were used to reflect the idea of 'Middle Class' men and women who would buy these baskets. The splotch of sauce on their faces injects a bit of fun and humor into these everyday products.

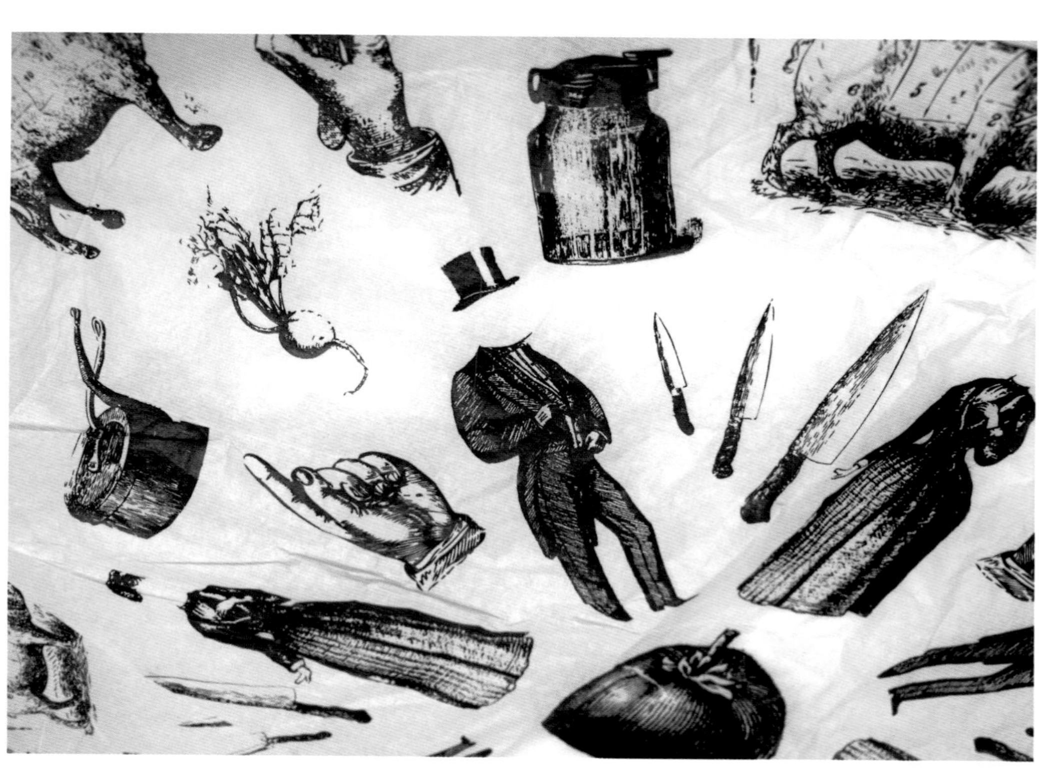

TOMMY
SAUCE
TOMMY
SAUCE

TOMMY
SAUCE
BROWN

Siete Pecados

agency **Sidecar**
Logroño Spain
art director Paco Valverde
designer Paco Valverde
www.sidecaronline.com

Design, naming and concept of a collection of wines inspired by the Seven Sins.
Lust, enveloped in a real woman's stocking, is evocative to the touch as soon as you pick up the bottle. Envy, coveting everything around it, hence its mirrored finish, reflecting and trapping its surroundings. Laziness, looking like it emerged from a cemetery of bottles where it had been sleeping for some undefined stretch of time, now covered in dust with its lettering tilted to the side, lying down, the way it prefers to be. On another bottle we find evidence of a display of unchecked Anger in the form of a burnt label... it couldn't be avoided. Pride, always looking down on the rest, now standing out with glittering Swarovski crystals. Gluttony was too much even for the cutlery, leaving it completely misshapen and useless from so much use and abuse. And finally, so that it lasts longer, not to be touched, a sturdy lock safeguards Greed.

siete
pecados
7
En
vi'
dia
INVIDIA
ENVY
ENVIE
NEID
RIOJA
DENOMINACIÓN DE ORIGEN CALIFICADA
siete
pecados
7
Gu
la,
RIOJA

RIOJA
Per
ez
za'
siete
pecados
7

Pecados

siete pecados
RIOJA
DENOMINACIÓN DE ORIGEN CALIFICADA
ira'
siete pecados
RIOJA
DENOMINACIÓN DE ORIGEN CALIFICADA
siete pecados
RIOJA
DENOMINACIÓN DE ORIGEN CALIFICADA
RIOJA

siete pecados
7
INVIDIA
ENVY
ENVIE
NEID
siete pecados
7
siete pecados
7
GULA
GLUTTONY
VOLLEREI
RIOJA
siete pecados
7
RIOJA

agency **Lavernia & Cienfuegos**
Valencia Spain
www.lavernia-cienfuegos.com

Ego aims to connect with a modern audience, concerned about their appearance, for a sophisticated elegance. The faceted glass pack has been painted in matte silver, such that the volume of the piece is solid and clearly defined. Ego uses a visual language that is direct and at the same time, refined.
The logo has been dealt with with equal strength, starting with a Didot typeface in which the characteristics of the letter 'g' have been enlarged so that in context, three letters together form a single entity, creating more personality.

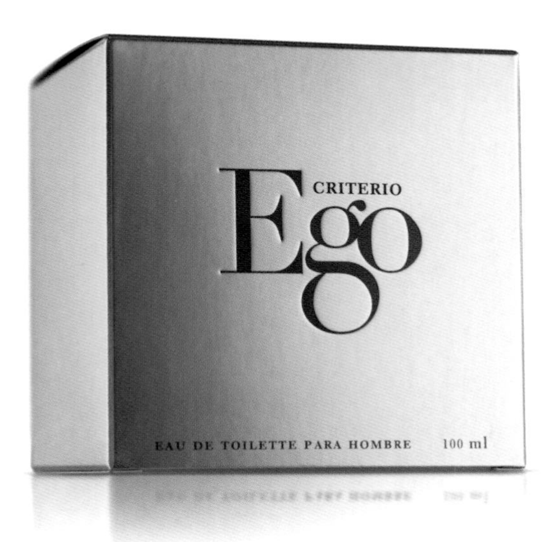

studio **Andreu Zaragoza**
Barcelona Spain
designer Andreu Zaragoza
www.andreu-zaragoza.com

Label design for Coma wine produced in Capçanes (Tarragona). As the name suggests, we can read the dictionary description of this word with a new definition that says: "A good wine from Capçanes." After designing the Coma wine label I wanted to make a packaging project that experimented with curved forms in cardboard. I used only black and white cardboard, without printing.

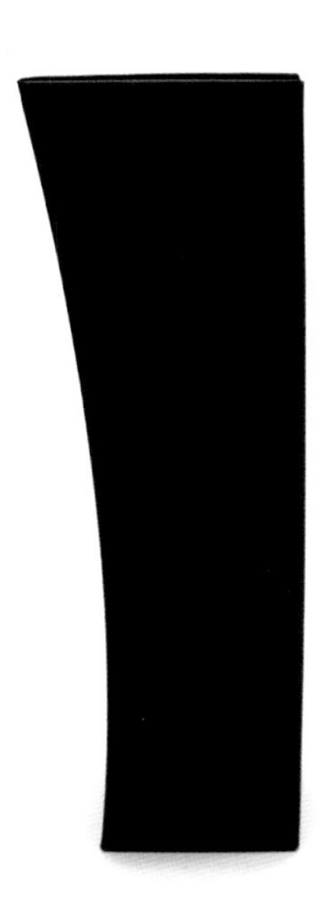

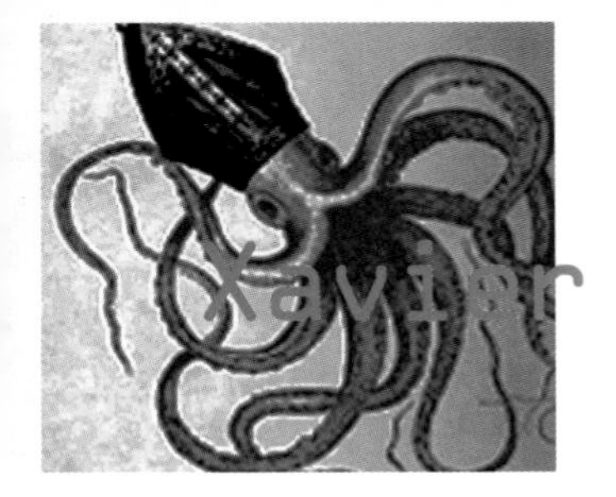

studio **Neumeister Strategic Design AB**
Stockholm Sweden
creative director Peter Neumeister
designer Per Torell
www.neumeister.se

Xavier Vignon is a world-class oenologist consultant who works for many of the leading estates in southern Rhône. He is known for creating wines of a truly unique character. But somehow, that distinctive styling didn't come across on the labels. Neumeister was contacted...
The unique character of every bottle catches the eye. If the wine has such a personal expression, it should be shown on every 'cuvee.'
The bottles were made to be really exceptional, one of a kind, using defining features: without ever losing the high quality wine ambience.

Xavier
Xavier
VACQUEYRAS
XAVIER VIGNON
Xavier
Xavier
CUVÉE ANONYME
CHÂTEAUNEUF DU PAPE
XAVIER VIGNON
Xavier
CHÂTEAUNEUF DU PAPE
XAVIER VIGNON
Xavier
Xavier
MERLOT
XAVIER VIGNON
Xavier
Xavier
VENTOUX
XAVIER VIGNON

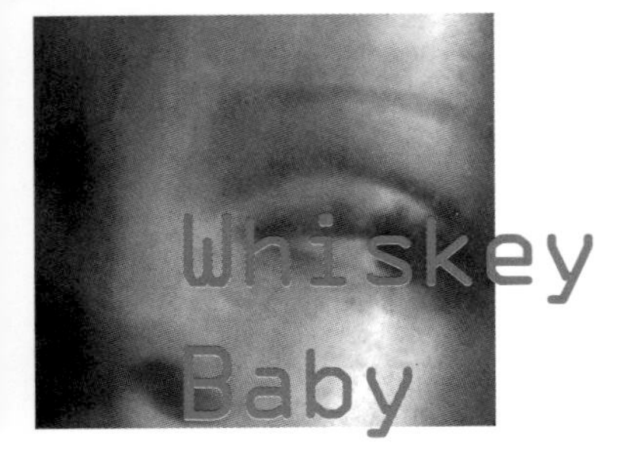

studio **Bayley Design**
London UK
designer William Suckling
www.behance.net/bayleydesign

Greta Garbo was a great actress of her era, however, it wasn't until she played the title role in Anna Christie (1930), that she said her first words on film. "Gimme a whiskey, ginger ale on the side and don't be stingy baby!" She said it with confidence, yet held back a relaxed attitude towards drinking the darker spirit. She was woman, why couldn't she drink whiskey?

The first challenge for this brief was to find ground on which it could stand. All whiskies have a long deep heritage and history. Making a modern whiskey meant not having that history. Applying the knowledge of Garbo's saying to a new whiskey, gave this new whiskey a history. The typography and label pay homage to films of Garbo's day, whilst keeping a fresh approach. The shape of the label derives from her ever curling and flowing hair. The shape also allows you to see to the back of the label where there is an image of Garbo.

Chilli

Dublin Ireland
designer Steve Simpson
www.stevesimpson.com

Mic's Chilli was a start-up when they approached me in 2010 to design their labels for their Inferno range of hot sauces. I wanted to approach the project from an illustration angle, making the design fit around the illustrations. The inspiration for the Sweet Chilli label designs came from traditional Chinese art, thus reflecting the origin of the sauce.

Boris Ice Tea

agency **lg2boutique**
Montréal Canada
creative director Claude Auchu
designers Caroline Reumont, Andrée Rouette, David Kessous
account services Catherine Lanctôt, Julie Bégin
strategy Marc-André Fafard
www.lg2boutique.com

Every flavor has a unique personality that is reflected in the packaging for this new family of alcoholic iced teas by Boris. The first one, lemon, is reminiscent of an eccentric character from England with a classic look, inspired by tea. The second, peach, highlights the feminine side and features a woman with flowing locks reminding us of the fuzz on the fruit. Each package is rendered in a fluorescent color palette based on the attributes of each fruit.

Boris
ICE
TEA
5,6%
alc./vol.
473 mL
ALCOHOLIC
MALT
BEVERAGE
WITH TEA
AND LEMON
TASTE

Boris Ice Tea

agency **lg2boutique**
Montréal Canada
creative director Claude Auchu
designers Caroline Reumont, Andrée Rouette, David Kessous
account services Catherine Lanctôt, Julie Bégin
strategy Marc-André Fafard
www.lg2boutique.com

Boris
ICE
TEA
5,6%
alc./vol.
473 mL
ALCOHOLIC
MALT
BEVERAGE
WITH TEA
AND PEACH
TASTE
Boris
THÉ GLACÉ
ICE TEA
THÉ
GLACÉ
Boris
ICE
TEA
5,6%
6X

Naked King

agency **Beetroot Design Group**
Thessaloniki Greece
designers Vagelis Liakos, Alexis Nikou & Yiannis Charalambopoulos
www.beetroot.gr

A new winemaker contacted us in order to provide a visual identity for his new wine. The wine is of extremely high quality, but free from any filtering or chemical treatments. So, it sits there exposed, without any chemical aid for the consumer to taste it. "You could say that it was the king of wines ...but a sincere one" the client said. That reminded us of the Hans Christian Andersen tale "The Emperor's Clothes" where the king sits naked and exposed, for everyone to see who he really was. So, we decided to name the wine NAKED KING. In order to visually present that, we designed the label on the bottle to be shaped like a crown and embossed the title NAKED KING, instead of simply printing it. We wanted the label to appear as if it were naked, with nothing but a crown on its head. For the wooden box, we told the tale in silk print and large type font, like in a children's book, and highlighted the words "The king is naked."

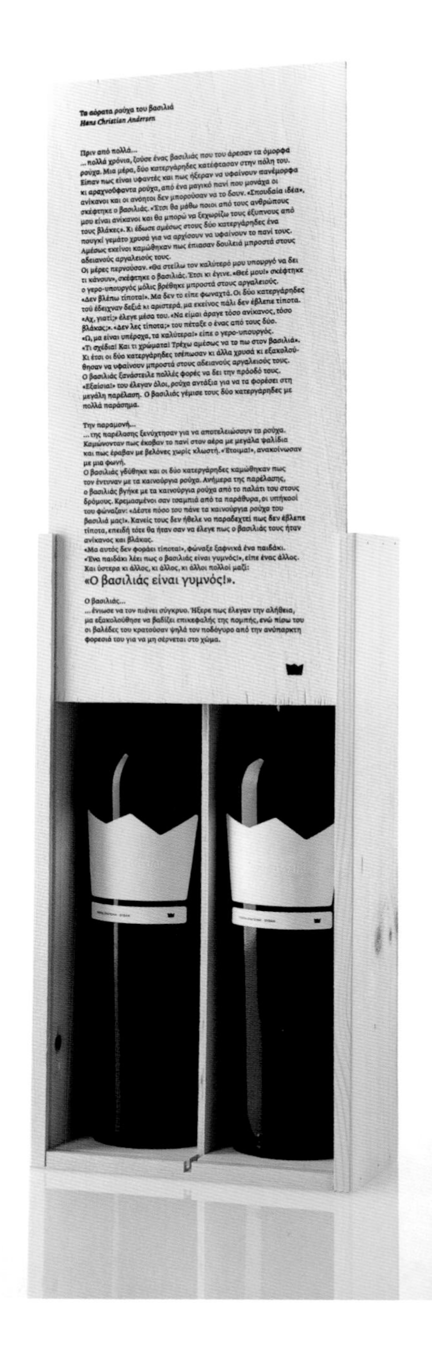

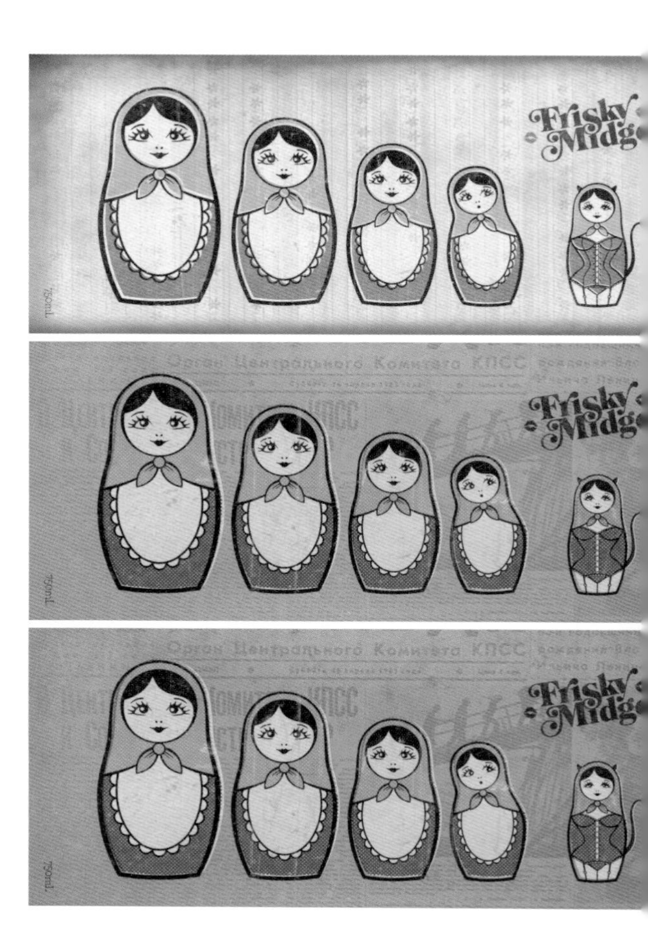

agency **Studio Lost & Found**
Perth Australia
designer & illustrator Daniel McKeating
photographer Daniel McKeating
www.studiolostandfound.com

Frisky Midget is the first in a series of quirky, tongue-in-cheek brands to be released by Killinchy Wines. The large wrap-around label is a throwback to Communist Russia. As you turn the bottle, a series of traditional Matryoshka dolls are revealed, with the final doll exposed as the saucy, non-conformist - Frisky Midget. This cheeky character is a celebration of individuality and the freedom of self-expression.
Printed four color process on Fasson Estate 8 premium uncoated paper.

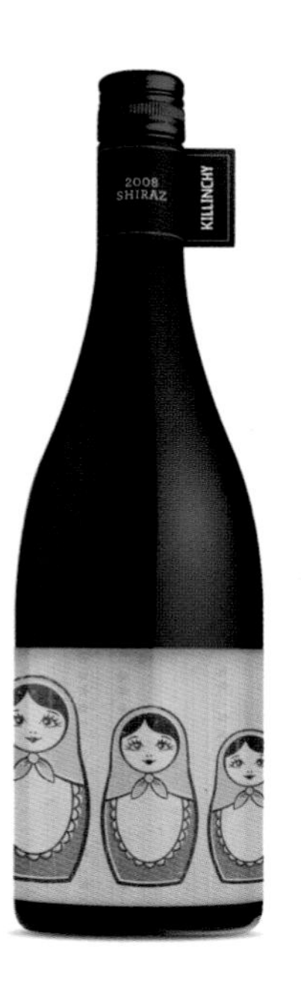

363009 093002
POKE FRISKY MIDGET
ON FACEBOOK!
www.friskymidget.com
Frisky
Midget

363009 093002
POKE FRISKY MIDGET
ON FACEBOOK!
www.friskymidget.com
КПСС
Frisky
Midget

363009 093002
POKE FRISKY MIDGET
ON FACEBOOK!
www.friskymidget.com
КПСС
Frisky
Midget

Build Your Own

agency **The Creative Method**
Sydney Australia
creative director Tony Ibbotson
designer Andi Yanto
artworker Greg Coles
www.thecreativemethod.com

The aim was to create a unique gift to give our clients at Christmas, while also acting as a new business introduction. The label needed to remind them of who we are and the long hours that we put into our work. It needed to feature all of our staff, reflect our creativity and our sense of humor.
We printed 5000 labels we created on our own and obtained high quality cleanskin wines. Each label was based on one staff member. It included a number of their facial features and the client is encouraged to BYO - Build Their Own - faces with features.
The wine and the label are the perfect substitute for when our team cannot be there.

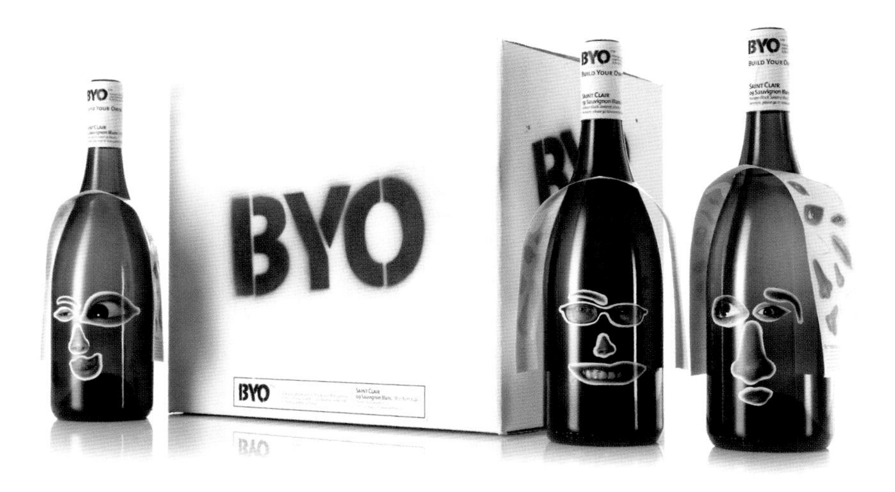
BYO
SAINT CLAIR

BYO
TCM
Sauvignon Blanc

Beervana

studio **Neumeister Strategic Design AB**
Stockholm Sweden
creative director Henrik Hallberg
designer Lachlan Bullock
www.neumeister.se

Brutal Brewing wanted to develop a limited edition summer beer – a light brew with a slightly lower alcohol content. A concept strong enough to make it stand out amongst other lagers during the busy summer season – sharp, humoristic and accessible. In search for the existential truth of craft beer we experimented, to free our minds.
With a pillow-soft and almost-in-motion typography, we finally reached true Beervana. A beer perfectly suited for harmonious gatherings in the sun.

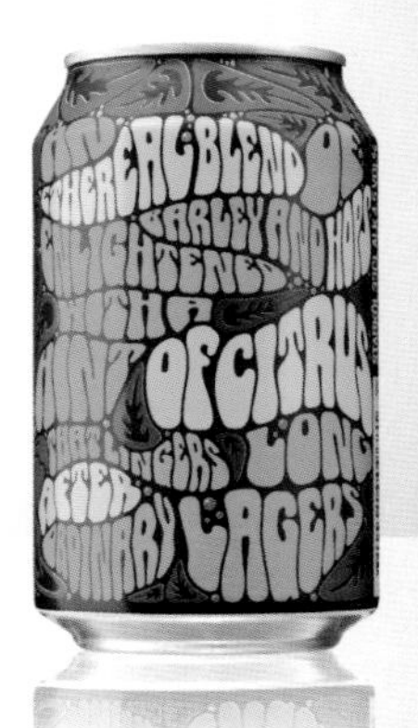

STARKÖL
BEERVANA
ALK 4,5% VOL
BRUTAL BREWING
BEERVANA
24 X 33 CL
ALK 4,5% VOL
BRUTAL BREWING

Beer Collection
Moscow Russia
designers Galima Akhmetzyanova & Pavla Chuykina
3D visualisation Pavel Gubin
www.cargocollective.com/hellogalima
www.behance.net/pavla

Hunting is one of the oldest human instincts. Do not restrain yourself. Let's go into the wild. Ignite your hunting passion and feel the rush of adrenaline. Track down prey. Get your trophy.
The idea started out as a joke about hunting trophies, but we decided to develop it into something more.
The result? Four labels with different animal characters and a box shaped as a forest. In addition, we added little custom details like bullet holes, a bullseye mark on the bottom of the bottle and a rifle-opener in the box!

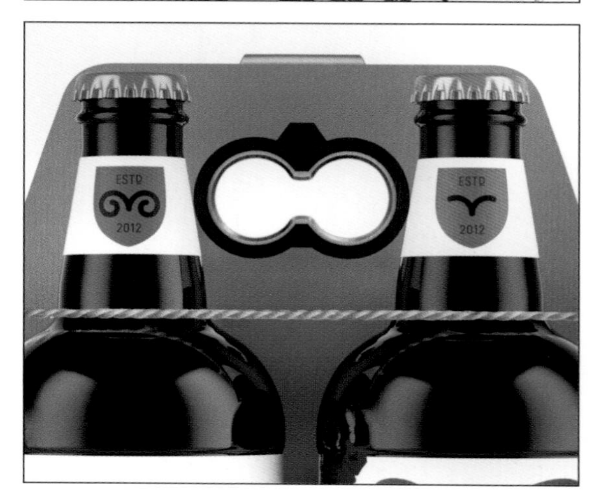

DEER
TROPHY
5% alc • BEER & BLUEBERRY • 0.5L

SHEEP
TROPHY
BEER & HERBALS
CONTAINS: water, malt, hops and cloudberry
CALORIES: 150 cCal • Carbohydrate 13g • 4.7%
Vikings believed that a giant goat whose udders provided an endless supply of beer was waiting for them in Valhalla which was vikings' heaven.
ENJOY IN MODERATION
Crafted according to Russian beer-making traditions.
KEEP REFRIGERATED • Best served at 8-12°C
WARNING
ALCOHOL REDUCES DRIVING ABILITY
5% alc
0.5 liters
4 670001 490627
ESTD
2012
GOAT
TROPHY
5% alc • BEER & CLOUDBERRY
ESTD
2012
YAK
TROPHY
5% alc • BEER & RASPBERRY • 0.5L

ESTD
2012
SHEEP
TROPHY
5% alc • BEER & HERBALS • 0.5L
ESTD
2012
GOAT
TROPHY
5% alc • BEER & CLOUDBERRY • 0.5L
ESTD
2012
TROPHY
5% alc • BEER & BLUEBERRY • 0.5L
ESTD
2012
YAK
TROPHY
5% alc • BEER & RASPBERRY • 0.5L

Libalis

agency **Moruba**
Logroño Spain
designer Javier Euba & Daniel Morales
illustrator Brosmind
www.moruba.es

Libalis is experiencing a big bang. Five years after its release, the wine by Maetierra has a new start, a reboot, a renewed commitment to surprise and pleasure. To express the moment, a visual explosion was created where the wildest situations coexist with references to episodes of the short and intense life of the brand. This is the start of the Libalis big bang.

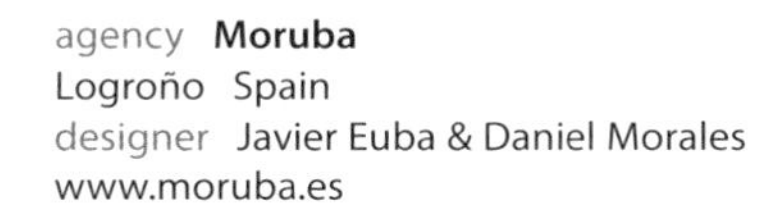

agency **Moruba**
Logroño Spain
designer Javier Euba & Daniel Morales
www.moruba.es

A town's barber shop is the center of the world. Here, mundane chat mingles with ancestral wisdom. Here, sometimes, something important begins. This is the wine of the barber.

A Couple Of Drops

agency **Beetroot Design Group**
Thessaloniki Greece
designers Vagelis Liakos, Alexis Nikou & Yiannis Charalambopoulos
www.beetroot.gr

Rich, velvety flavor, fruity and earthy aroma and a luxurious golden color with subtle green tinges.
We welcome you to experience a couple of drops extra virgin olive oil. Extra virgin olive oil from the Peloponnesian land of Kalamata. Extra virgin olive oil as it should be! Olive oil in Greece is sacred: a "divine gift." It is our history, nutrition, nourishment and ritual. It is the nutritious "gold" that kept the Greek civilization's flame alive throughout its long history: from the Ancient times, through the Byzantine Empire and until now. For us Greeks, 'oil' always means 'olive oil' and it is always present in a large bottle, holding a sacred place, within our kitchens. It is one with the land, the sun and the sea.

A COUPLE
OF DROPS
of GREEK
EXTRA
VIRGIN
OLIVE
OIL
500 ML
PRODUCT OF GREECE
Acidity ≤ 0.5%
We welcome you to experience a couple of drops of extra virgin olive oil, produced in the Peloponnesian land of Kalamata. Rich, velvety flavor, fruity and earthy aroma and a luxurious golden color with a subtle green tinges...
The extra virgin olive oil as it should be!
OLIVE OIL
www.agreekculture.gr

A COUPLE
OF DROPS
of GREEK
EXTRA
VIRGIN
OLIVE
OIL
500 ML
PRODUCT OF GREECE
Acidity ≤ 0.5%
Superior category olive oil obtained directly from olives and solely by mechanical means.
Nutritional Facts per 100ml
824 Kcal / 3389kj
Energy 0g
Proteins 91.5g
Total Fat 6.5g
- Polyunsaturates 72g
- Monounsaturates 13g
- Saturates 0mg
Cholesterol
Store in a cool place, away from light.
Quality characteristics
Acidity ≤ 0.5%
Petroxides ≤ 20meq O2/kg
Waxes (ppm) ≤ 250mg/kg
Ultraviolet absorption K232 ≤ 2.50
K270 ≤ 0.22, DK ≤ 0.01
Cold extraction at a temperature below 27°C
Certified by ELOT EN ISO 900:2000
Packed by Sarantos A. Polizois,
24001, Kremmidia, Messinia, Greece
EI 40057
5 214000 052029

Dublin Ireland
designer Steve Simpson
www.stevesimpson.com

These are the first 4 labels for Mic's Chilli's Inferno range of Hot Sauces. The range has four strengths - 4 chillies Extreme, is the hottest down to 1 chilli JNR, the mildest. Inspiration for the labels was taken from the Mexican holiday, Day of the Dead. The barcodes are also illustrated in the same theme.

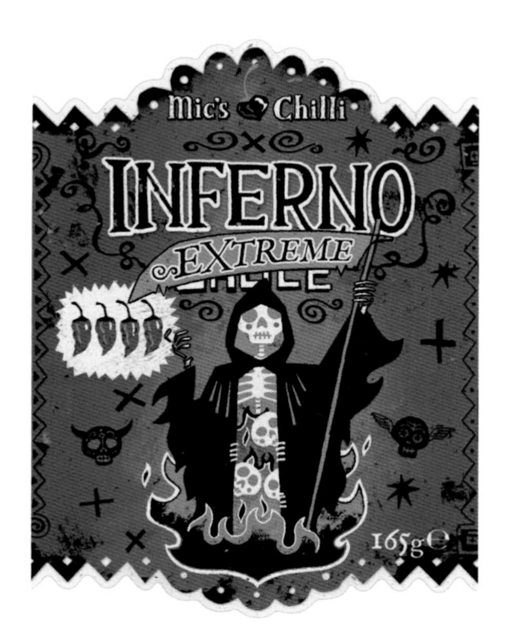

INFERNO
INFERNO
INFERNO
INFERNO

Mic's Chilli
INFERNO
EXTREME
Inferno Extreme
Judgement day for chilli lovers
Warning: Even the most hardcore chilli fans will find this blazing hot! With 12 Haberno Chillies, this really isn't for the uninitiated!
Despite it's name, Inferno Extreme is a natural beauty & has no preservatives or flavourings inside.
The Inferno Formula
Made by Lord Mic in Kilcoole, Co. Wicklow, Ireland.
Visit us at www.MicsChilli.ie
It's what's inside that counts...
12 {yes, twelve!} Habanero Chillies (40%), Water, Distilled Vinegar, Lemon Juice, Spices, Salt, Garlic, Xanthan Gum.
Nutrition values per 100g
Energy 147kJ/35kcal Protein 1g Carbs 8g Fat 1g
Storage tips
Refrigerate & consume within 3 months.
5391519890301
Best before: See base of bottle
165g℮

Mic's Chilli
INFERNO
SAUCE
165g℮

Mic's Chilli
INFERNO
SAUCE
lite
165g℮

studio **Yevgeny Razumov**
Moscow Russia
designer Yevgeny Razumov
www.behance.net/razumov

In Italian, divino means divine and vino is wine. Padre Divino is a dessert red wine recommended for use on Catholic holidays. It also serves the purpose of one of the basic elements of receiving the Eucharist in Catholicism.

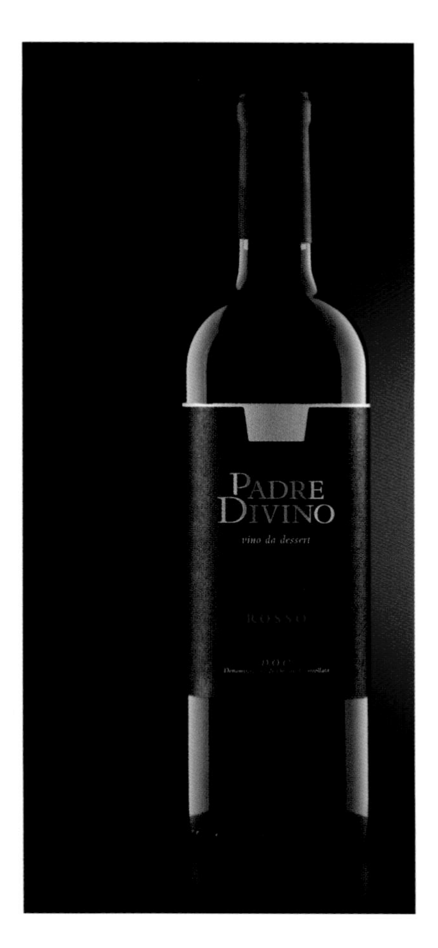

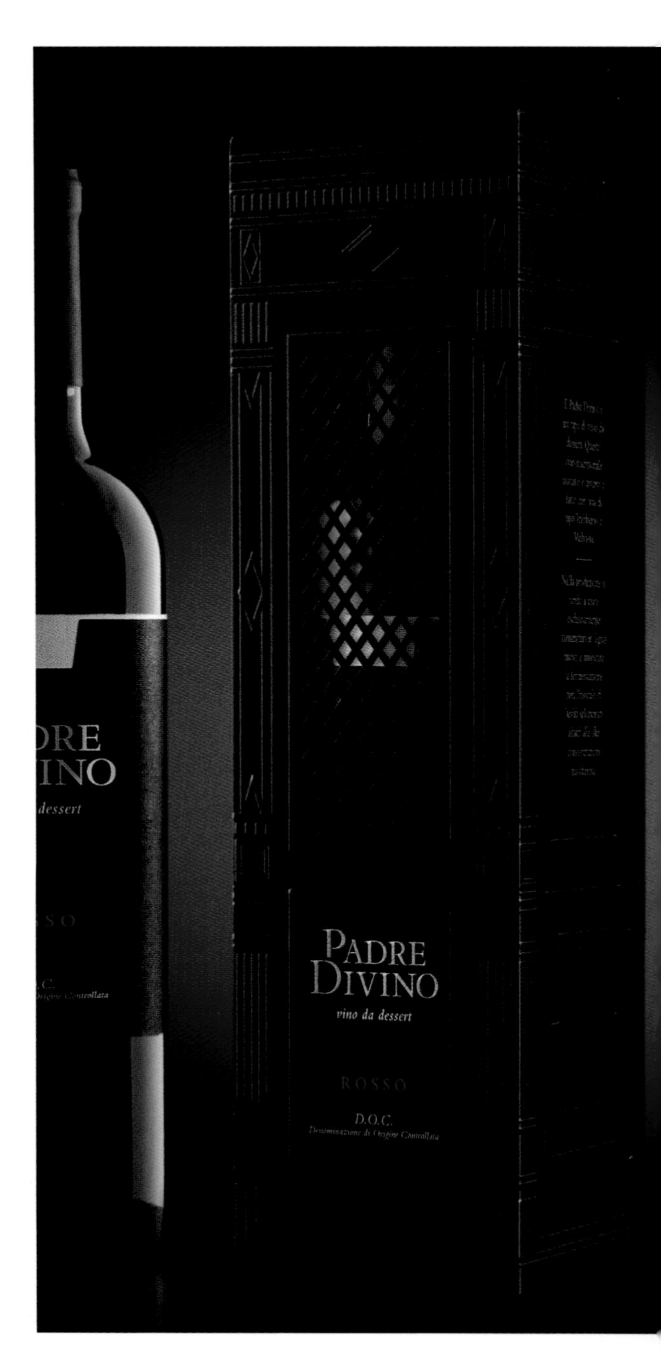

Chocolate glue

agency **Kolle Rebbe / KOREFE**
Hamburg Germany
creative director Katrin Oeding
designer / illustrator Reginald Wagner
photography Ulrike Kirmse
www.korefe.de

Chocolate Glue is bottled in a classic wood glue bottle and sticks together with any type of bread. When the bottle is turned upside down, the typographic chocolate monster rolls his eyes. Chocolate Glue is available in chocolate and chocolate-biscuit flavors.
The Deli Garage is a food label that sells delicatessen products with the design and practicality of everyday DIY and garage items.

agency **Moxie Sozo**
Boulder, CO USA
designer Nate Dyer
www.moxiesozo.com

Targeting parents, young adults, children and health-conscious women, iWellness brought to market Picabi; a line of sparkling juices intended to provide delicious and nutritious benefits to the masses by combining premium ingredients with the flavor and aroma of gourmet juices. Picabi's launch pegs iWellness as the first and only company offering consumers 100% sparkling juice with significant health benefits.
In addition to creating Picabi's logo, Moxie Sozo also developed packaging design that generated intense visual appeal, while our lively and youthful exhibit banner and print collateral drew new customers to the Picabi product line. By giving fruits a face, Moxie Sozo was able to bring the Picabi drink flavors to life.

picabi
SPARKLING COCONUT WATER
8.4 FL. OZ. (250 mL)

picabi
100% SPARKLING GRAPE JUICE
8.4 FL. OZ. (250 mL)

picabi

Tapas wine Collection

studio **Eduardo d Fraile**
Murcia Spain
designer Eduardo d Fraile
www.eduardodfraile.com

The Tapas Wine Collection is a brand that was created to export good Spanish wines at an affordable price. The graphics were inspired by the traditional elements of bars in Spain. Spanish classics often featured on bar menus are written on the bottle, giving the appearence of chalk written on a wall. The waiter shouts out to the cook: "a ham! a tortilla!" The patron, after eating his usual meal, cleans his teeth with a wooden stick.

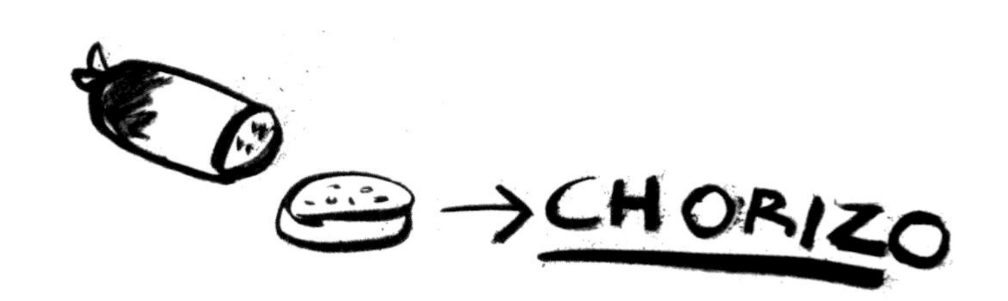
→CHORIZO

THE
TAPAS
WINE
COLLECTION
VINO TINTO
2007
OLIVE
OIL
BREAD
JAMÓN
TORTILLA
CHORIZO
PALILLO!
UNA COPITA
DE VINO TINTO,
UN PINCHO DE
TORTILLA ...

Martini Art

studio **Istratova Alexandra**
Moscow Russia
designer Istratova Alexandra
www.behance.net/Sasha-Tyla

Limited Edition for Martini.
New Year - a time of magic, when you can wear a mask and play...
A series of dresses - strict, fun, feminine and flamboyant - satisfy every taste, grace every evening.
The mask becomes an accesory to the party and even when the drink is drunk, the mood of the martini lives on.

MARTINI
RISERVA VERNICE

agency **The Creative Method**
Sydney Australia
designer Tony Ibbotson
www.thecreativemethod.com

The brief was to develop a wine package for TCM's clients that would remind them of our ability to think differently. We created a label with immediate impact, that stood out, while also reminding them of the end of year party.
The name of the label is Holy Water and each staff member was photographed and brought to life in a traditional horror book style. The story follows the characters experimentation with Holy Water and the catestrophic results. Each bottle contains 2 pages, with a total of 6 in the set. Different varieties are illustrated with a different base and retro style color.

PAGE
3&4

PAGE
5&6

HOLY WATER
YOU ARE, MERE SECONDS FROM YOUR DOOM!
SQUIRT!
OH MY GOD!!!
2010 SAUVIGNON BLANC!!
THE FINEST SQUEEZED GRAPE JUICES GIVE BIRTH TO WINE THAT IS OUT OF THIS WORLD!
13.5%ABV 750mL
YOU MAY HAVE DESTROYED MY WORK BUT YOU'LL NOT ESCAPE ALIVE!
AS THE ARCHFOES DUEL TO THE DEATH, FROZEN LIMBS SLOWLY THAW ... RHEUMY EYE SOCKETS START TO CLEAR ...
... AND THE DEAD COME TO LIFE, OBLIVIOUS TO THE BRAWLING COMBATANTS ...
HA! YOUR EFFORTS ARE IN VANE. OH FERNANDA! LOOK AT HER! IS SHE NOT DIVINE!
HOLY WATER
BARON! I HUNGER! MUST! HAVE! BLOOD!
NO! I SHALL SLAKE MY THIRST ... ON YOU!
NO! WHAT ARE YOU DOING? I AM YOUR MASTER! OBEY ME!
OH MY GOD!!!
2010 PINOT NOIR!
THE FINEST SQUEEZED GRAPE JUICES GIVE BIRTH TO WINE THAT IS OUT OF THIS WORLD!
14%ABV 750mL
UNHAND ME, UNCLEAN CREATURES! YOU WILL LEARN TO DO AS YOU'RE TOLD!
RRARR! BARON - - NOT HURT - - WOMAN!
ACKK!! G-GASP!

Lolita

agency **Sidecar**
Logroño Spain
art director Paco Valverde
designer Paco Valverde
www.sidecaronline.com

Lolita is a premium extra virgin olive oil.
The design is based on the book by Vladimir Nabokov (Lolita), in which a mature teacher falls for a girl who is only 12 years old. The design tries to capture that inappropriate love through a sober and minimalist symbol. It is an inverted heart, which becomes an ass (he thinks he's in love, but is driven by a sick lust). The name was chosen because it contains the word Lolita, OLI (Olive), and is a true virgin. It is a game that aims to make the customer think.

Lolita.
"Era Lo, sencillamente Lo, por la mañana, un metro cuarenta y ocho de estatura con pies descalzos. Era Lola con pantalones. Era Dolly en la escuela. Era Dolores cuando firmaba. Pero en mis brazos era siempre Lolita."
"She was Lo, plain Lo, in the morning, standing four feet ten in one sock. She was Lola in slacks. She was Dolly at school. She was Dolores on the dotted line. But in my arms she was always Lolita."
Vladimir Nabokov
ACEITE
DE OLIVA
VIRGEN
EXTRA
ecológico
Lolita.

Vinos 365 Delhaize

agency **Lavernia & Cienfuegos**
Valencia Spain
www.lavernia-cienfuegos.com

This is a wine line that Belgian supermarket Delhaize offers under their own brand "365", which covers affordable everyday products.
We set as our starting point the homogeneity of the line and a communication in line with the spirit of the "365" brand.
The cork is a sign of humility. A little value object, often used for arts and crafts - a simple and easy to manipulate element - to play and create with. The cork is the element that unifies and personalizes the entire line.
The motif designed for each label refers to the country of origin. In the case of France, the different types of wine are characterized by a type of French cap, the "canotier", Napoleon's hat.

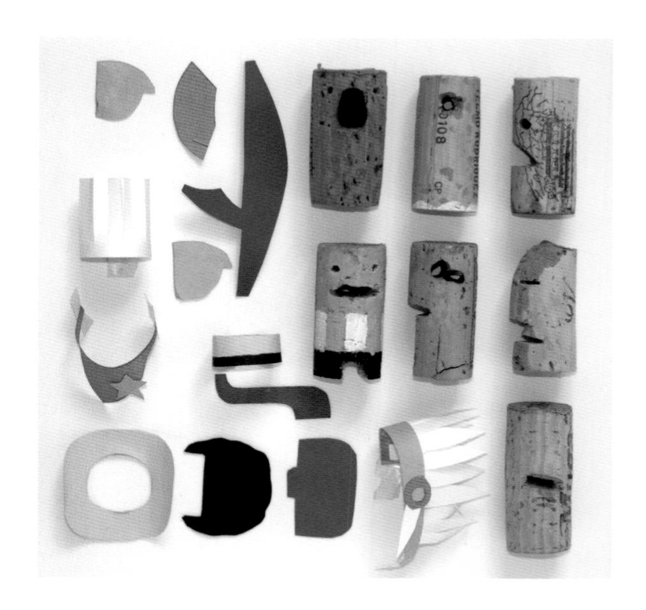

Coteaux d'Aix en Provence
Appellation d'Origine Contrôlée
Bergerac
Appellation d'Origine Contrôlée
Côtes du Rhône
Appellation d'Origine Contrôlée

Vinos 365 / Delhaize

agency **Lavernia & Cienfuegos**
Valencia Spain
www.lavernia-cienfuegos.com

Cabernet Sauvignon
Vino Tinto · Chile
365
Australian
White Wine · Chardonnay
365
Chardonnay
White Wine of California
365

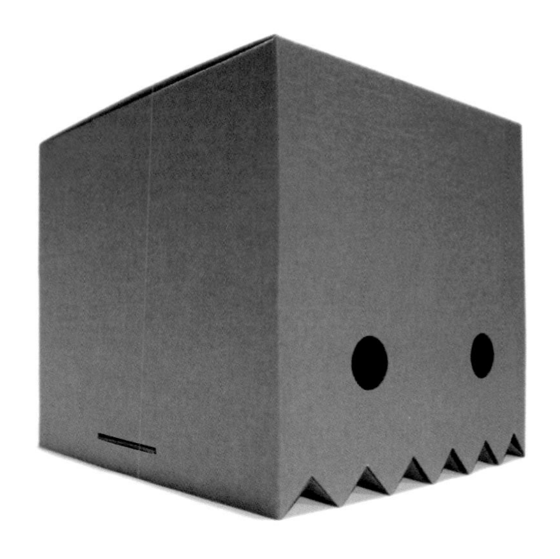

PIETRO
GALA
400 g
14.1 oz
linguini

PIETRO
GALA
400 g
14.1 oz
conchiglie

Packages for boxes and bags

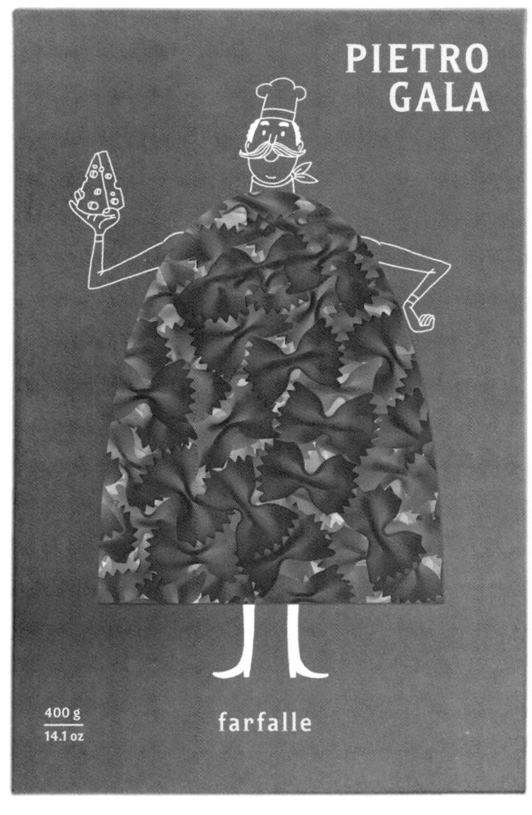

Delilah

studio **Lun Yau**
London UK
designer Lun Yau
www.lunyau.co.uk

The premise was to use the current Delilah branding and redesign the tea, coffee, voucher and bags as part of the "DIY Hamper."
Delilah is a delicatessen that sells selective foods from around the world. The food at Delilah speaks for itself - they are exotic, quality produce. Taking this concept, the packaging reflects the honest nature of Delilah's foods.
The 'food speaks for itself' approach lead to a literal and visual approach by the designers. The speech bubble is a marque, created to be consistently applied on Delilah's packaging. The gift hamper packaging shows speech bubbles that can be customised by the customer, which add a personalised gift element to their purchases. The packaging is inspired by chalkboards and give an organic and personal feel.
The packaging for tea and coffee used the same screen printed speech bubble with the Delilah logo, as if Delilah is a person speaking. Colored chalks are supplied within the hamper, so that the packaging can become a personalised gift.

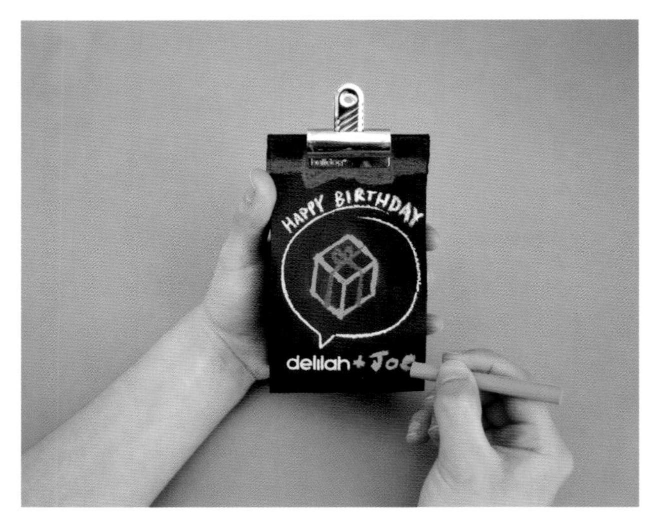

delilah
Coffee.
Tea.
delilah
delilah
£10
HAPPY BIRTHDAY
TO
delilah
delilah + Joe
MERRY
CHRISTMAS
xx
delilah

agency **Fresh Chicken**
St. Petersburg Russia
design Fresh Chicken
www.frch.ru

Pietro Gala is a new premium pasta brand, distinguished by handcrafted manufacturing and high quality ingredients. The Fresh Chicken agency developed the brand name and character and designed the production package. Pietro Gala is an Italian chief whose image now features on different kinds of pasta. The one-color print and cardboard texture emphasize the naturalness of the pasta and generate positive emotions.

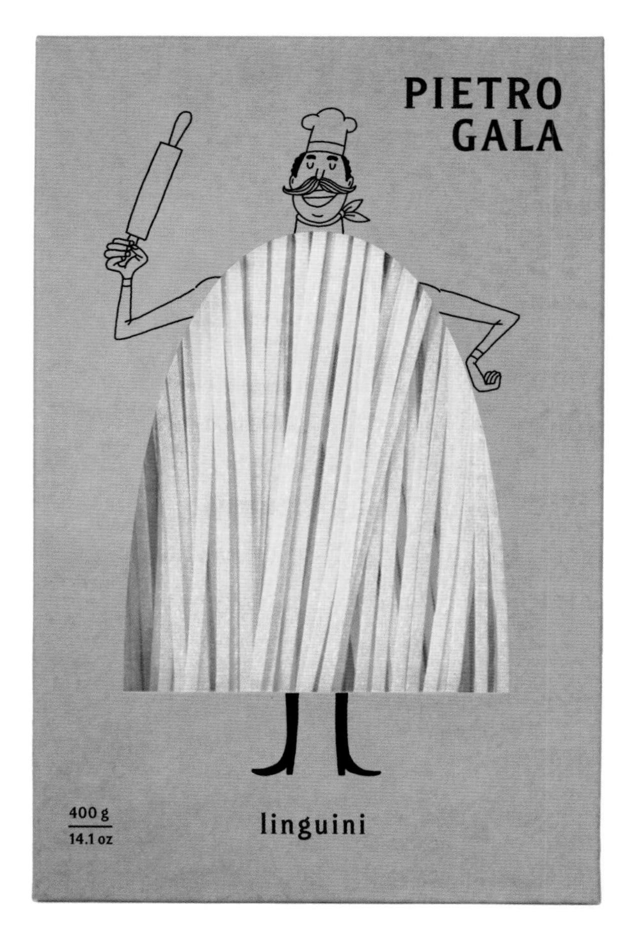

PIETRO
GALA
400 g
14.1 oz
linguini
PIETRO
GALA
400 g
14.1 oz
conchiglie

PIETRO
GALA
radiatori

PIETRO
GALA
400 g
14.1 oz
farfalle

Cookie Break

school **British Higher School of Art and Design**
Moscow Russia
designers Artem Maslov, Tatyana Rusalovskaya, Alya Lugovaya, Stas Semin

This packaging is for a new line of organic products. It is a healthy and complex snack-set: cookies, natural yogurt and homemade jam (kiwi, cherry or orange). All of the ingredients are organic and fresh. 3 cardboard tubes are placed on top of each other inside a shrink-wrapped container. The brand information is printed on the wrapper. The tubes contain only a photo of the content and a description of the product. When the customer is removing the wrapper, he or she is metaphorically becoming free from the chains of the urban hustle and is left alone with the product.

Healthy Snack
cherry jam+yogurt+cookies
12. 12 12:23
cherry jam

COOKIE BREAK
Healthy Snack

Cookie Break

school **British Higher School of Art and Design**
Moscow Russia
designers Artem Maslov, Tatyana Rusalovskaya, Alya Lugovaya, Stas Semin

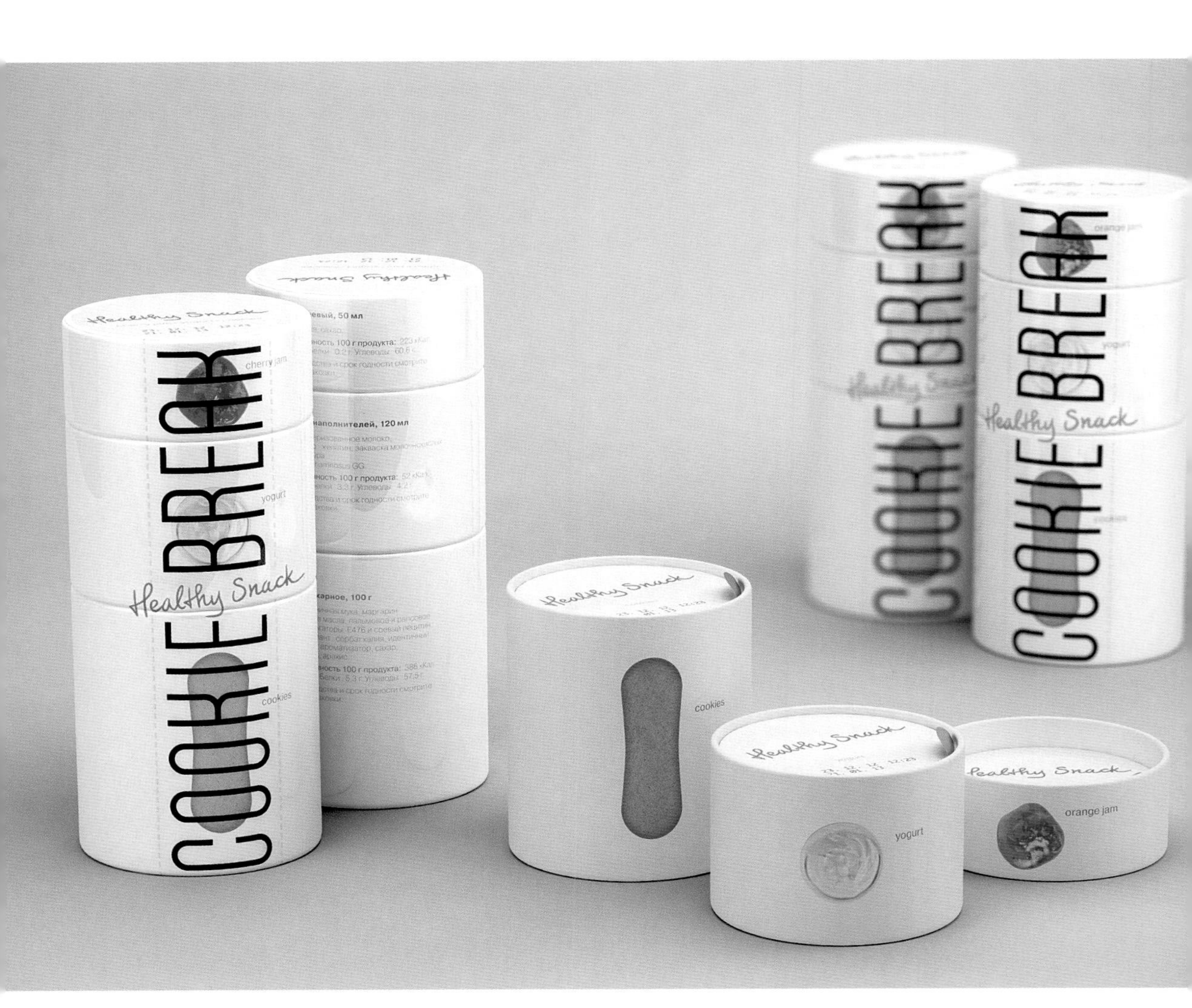

cookies
yogurt
cherry jam

COOKIE BREAK
Healthy Snack
COOKIE BREAK
Healthy Snack
yogurt
cookies
COOKIE BREAK
Healthy Snack

Silver Cross

agency **LOVE**
Manchester UK
www.lovecreative.com

LOVE's challenge was to marry the feel good factor of the brand's heritage with a new approach that spoke to parents and children and created a point of difference that separated Silver Cross from its competitors.
LOVE took the opportunity to project Silver Cross as a unique voice that reflected the reality of parenting, while creating empathy with the consumer. "Passionate About Parenting" was developed to make Silver Cross a unique 'parenting brand.' The tone of voice is real, honest and sometimes funny.

3D Accessories
Silver Cross 3D Accessories

Silver Cross 3D Pram System
Silver Cross 3D Pram System

Noodles

agency **Kolle Rebbe / KOREFE**
Hamburg Germany
creative director Katrin Oeding
designer/illustrator Reginald Wagner, Jan Hartwig
photography Ulrike Kirmse
www.korefe.de

Multi-Noodles, neatly packaged and sorted in screw boxes, should be part of any workshop dedicated to pleasure. Shaped like nails, screws, nuts and bolts, the noodles keep pasta dishes together and anything else that comes to mind in a creative kitchen.
The Deli Garage is a food label that sells delicatessen products with the design and practicality of everyday DIY and garage items.

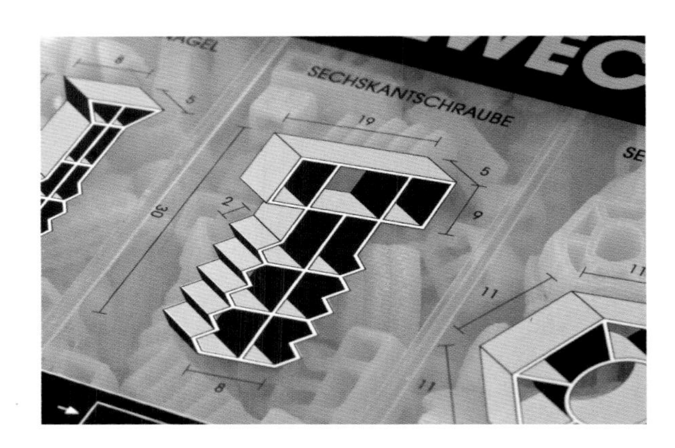

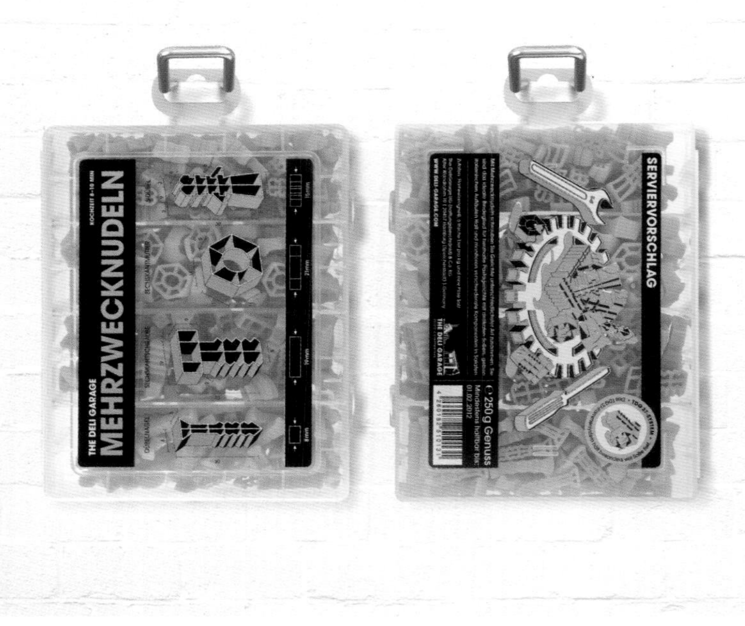

Lollitool

agency **Kolle Rebbe / KOREFE**
Hamburg Germany
creative director Katrin Oeding
designer/illustrator Reginald Wagner, Christine Knies
photography Ulrike Kirmse
www.korefe.de

Work could taste so sweet: with lollipop tools in a classic leather tool bag. The bag contains 6 lollipops in a variety of flavors, all shaped like screwdrivers. The handle is not the stick, but rather the lollipop.

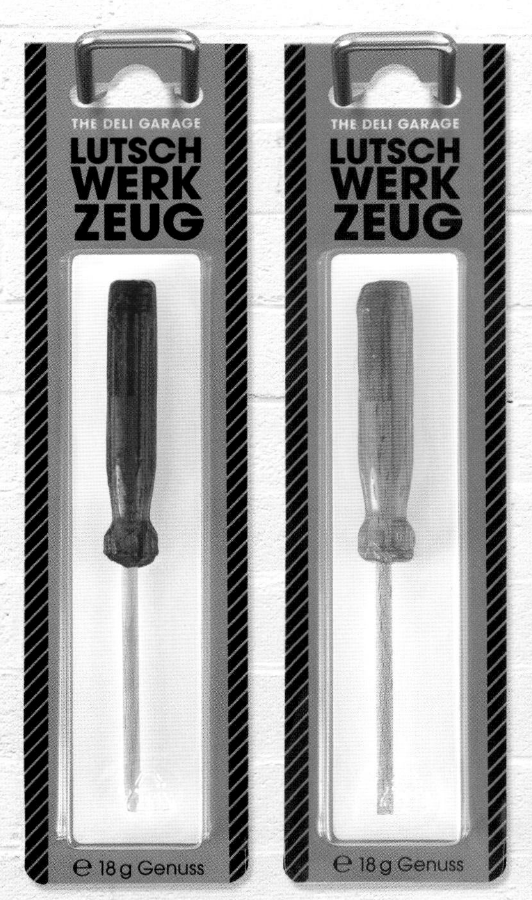

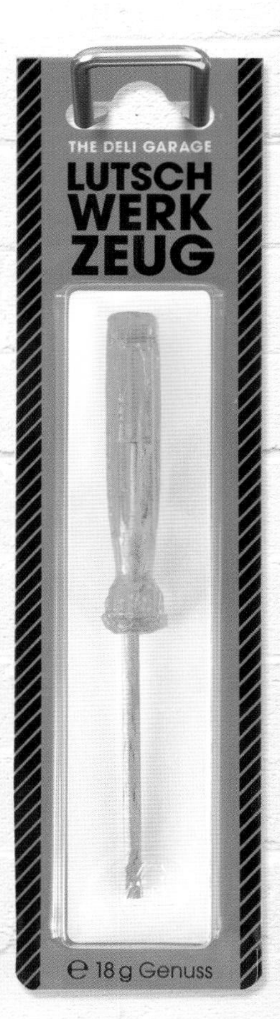

Cooking Good Taste

studio **Bessermachen Designstudio**
Copenhague Denmark
creative director Kristin Brandt
designer Kristin Brandt
www.bessermachen.com

In 2010, Eva Solo launched a series of cookware that combined the quality of professional cooking utensils with Eva Solo's distinctive and uncompromising design. A term was created, which united the two worlds.
The idea for the packaging design came from using raw materials as a tight design element, making use of their natural aesthetics, and putting it on the same footing as the XO products. The design let form and function meet in a culinary way!

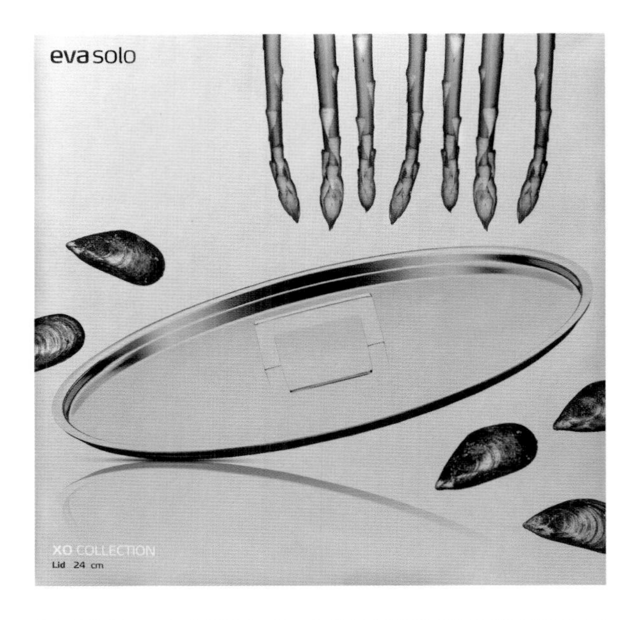

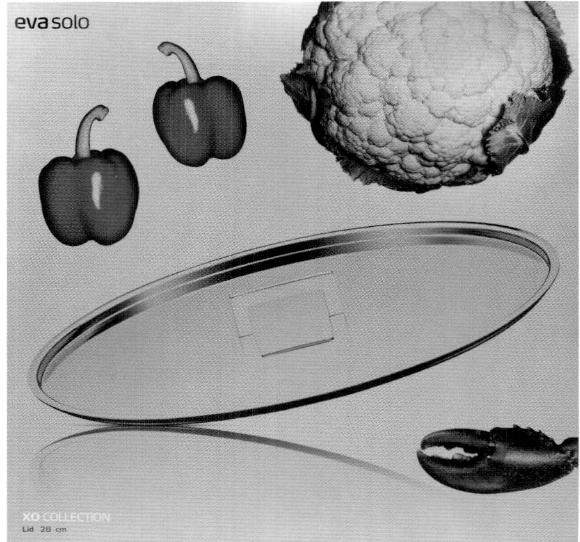

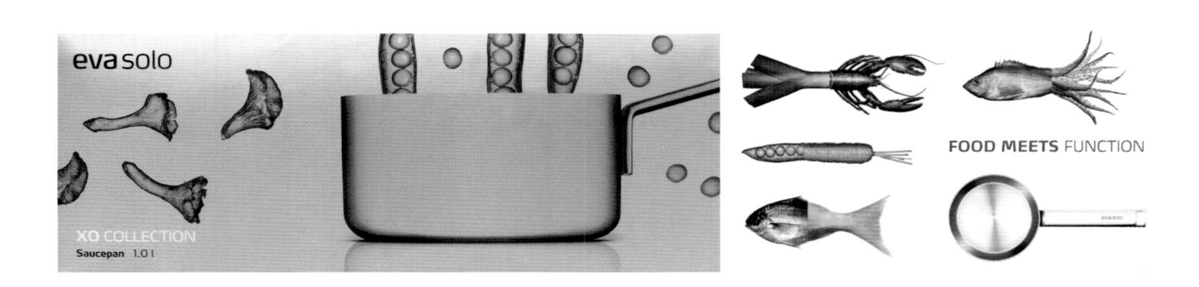

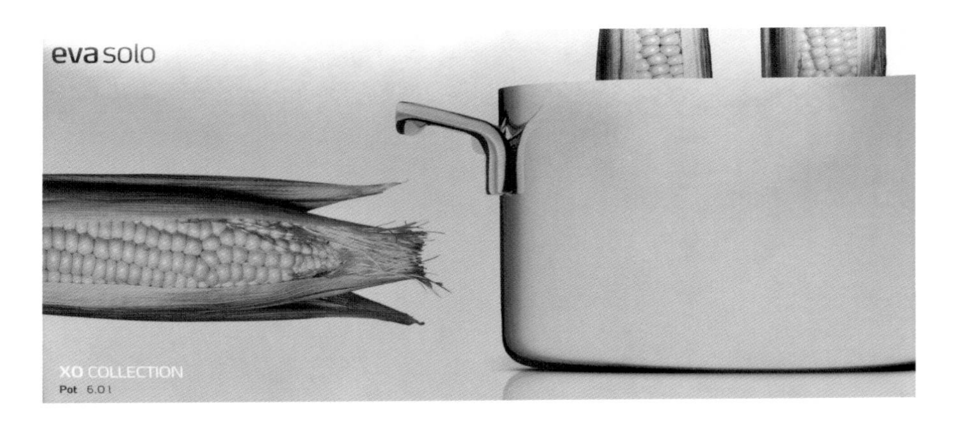
eva solo
XO COLLECTION
Pot 6.0 l

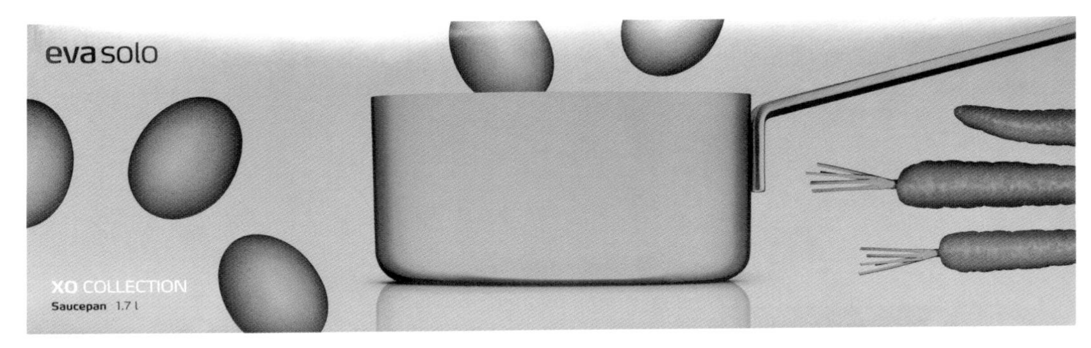
eva solo
XO COLLECTION
Saucepan 1.7 l

eva solo
eva solo

agency **Gworkshop Design**
Barcelona Spain
designers José Luis García Eguiguren,
Leonor Pinto, Catia Caerio
www.gworkshopdesign.com

The project: the development of a name, concept and proposal for packaging. The project should denote a great elegance and exclusivity, while also costing no more than seventeen euros.
The briefing describes our target audience: a secure woman, demanding, elegant, seductive with personality, taste and style. The element we used was smell; something that suggests something hidden, mysterious and secret.
The packaging evokes sensual aromas, an exclusive look and quality. "Minuit privé" means midnight, and the three names chosen for each product were: Desirée (desired), Seduisante (seductive) and Mysterieuse (mysterious).
The bottle was developed based on modern design, but also evokes something of the past. The body of the bottle takes us back to a more classic perfume style and the perfume cover gives us a modern touch, while linking the concept of smoke forming from liquid.

HELENA HESS
MINUIT PRIVÉ

HELENA HESS
MINUIT PRIVÉ
- SEDUISANTE -

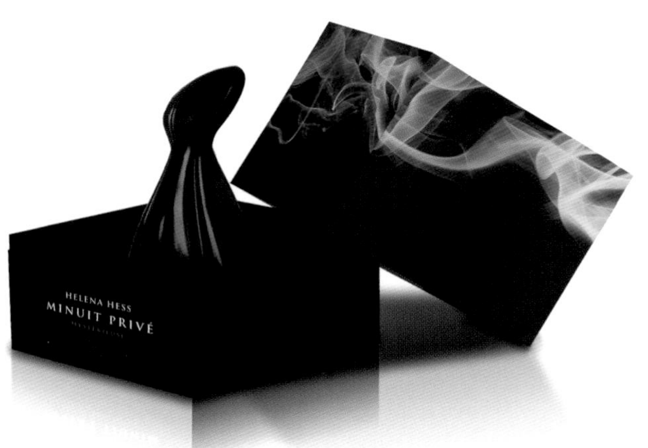
HELENA HESS
MINUIT PRIVÉ

HELENA HESS
MINUIT PRIVÉ
- SEDUISANTE -

HELENA HESS
MINUIT PRIVÉ

Hatziyiannakis dragees & pebbles

agency **Mousegraphics**
Athens Greece
art director Greg Tsaknakis
designer Kostas Vlachakis
illustrator Ioanna Papaioannou
photographer Dimitris Poupalos, George Telis
www.mousegraphics.gr

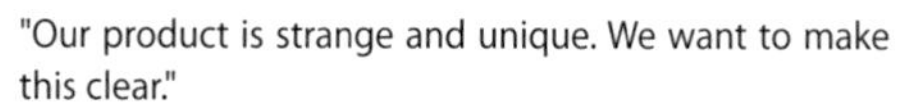

"Our product is strange and unique. We want to make this clear."
The target audience: bold and brave consumers, with a sweet tooth and a flair for discoveries. Intellectuals of the gourmet type. The design: to paraphrase the famous Marshall McLuhan expression, "the medium is the content" in this packaging extravaganza. The symbiotic relationship of medium and message (of which the media guru preached), is here translated into the eccentric coupling of package-product. An 'outer-inner' game of illusions is played for the eyes of the consumer. The paradox of a sweet, edible, even appetizing pebble. The beauty of an open crop, with its shockingly realistic flesh; fake cherries which can fool birds into coming to nibble on them like in the ancient paintings of Zeuxis, rocky-tasty formations: all these are mind treats we prepared for the consumer within a heightened - reality designed - environment.

HATZIYIANNAKIS

Hatziyiannakis dragees & pebbles

agency **Mousegraphics**
Athens Greece
art director Greg Tsaknakis
designer Kostas Vlachakis
illustrator Ioanna Papaioannou
photographer Dimitris Poupalos, George Telis
www.mousegraphics.gr

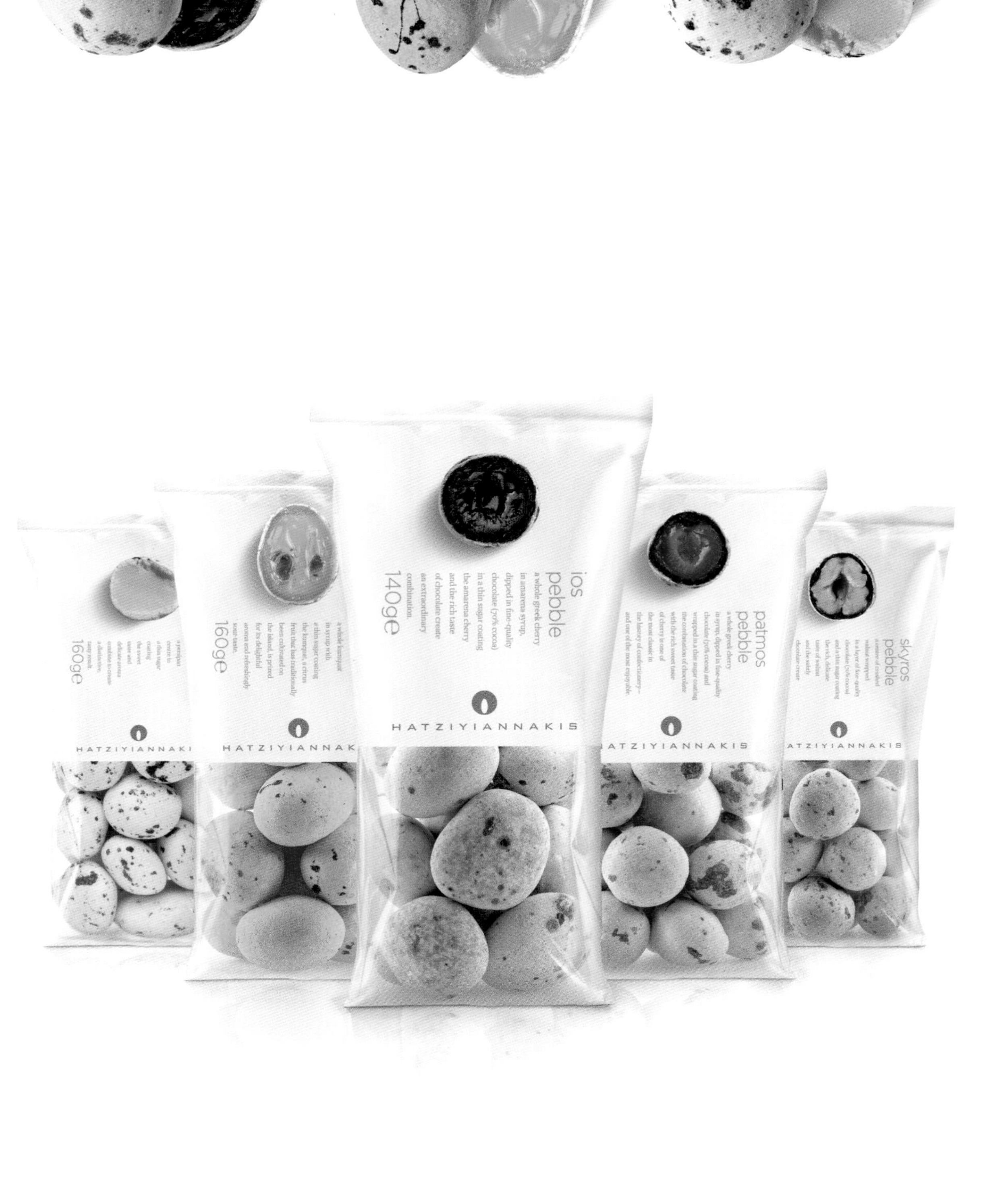
160ge
HATZIYIANNAKI
160ge
HATZIYIANNAK
ios
pebble
140ge
HATZIYIANNAKIS
patmos
pebble
ATZIYIANNAKIS
skyros
pebble
ATZIYIANNAKIS

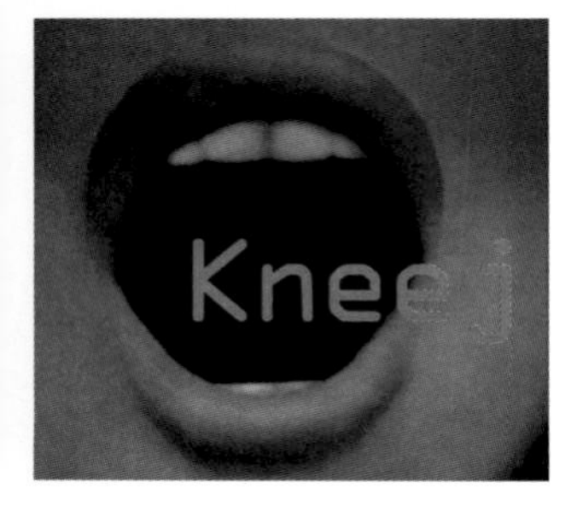

agency **Happy Creative Service**
Bangalore India
creative director Praveen Das, Kartik Iyer
art director Praveen Das, Pradeep Kumar
copywriter Sanaa Abdussamad
illustrator Rishidev
designer Praveen Das
www.thinkhappy.biz

The task at hand was to come up with an in-store promotion to run alongside the 'Sex sells, unfortunately we sell jeans' campaign to help increase sales. So we created the Knee.J, a spoof sex toy of sorts to be given away to all customers that ran a bill above $150. One day after running posters and e-mailers, the stores saw an instant increase in walk-ins. Customers were eager to get their hands on the Knee.J. What was meant to be a one-store promotion, got enough publicity to manufacture more pieces and take them across all stores. Sales picked up by 40% and customers were spotted kneeling for more.

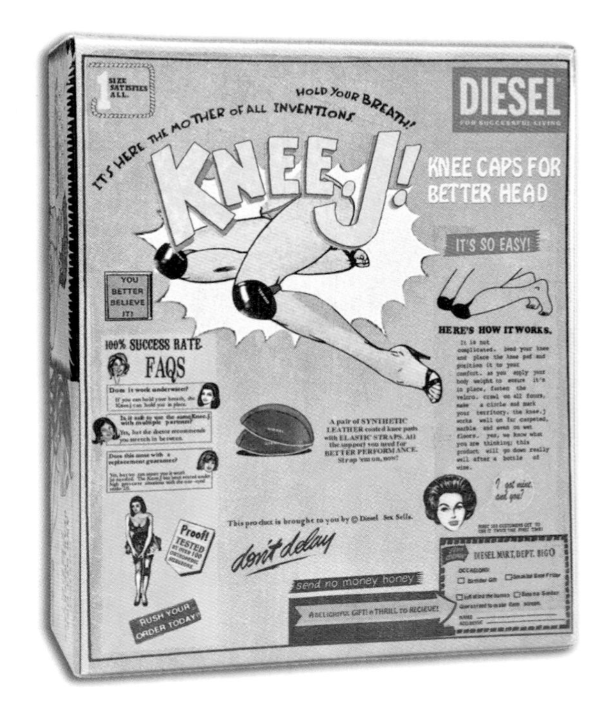

Peperman

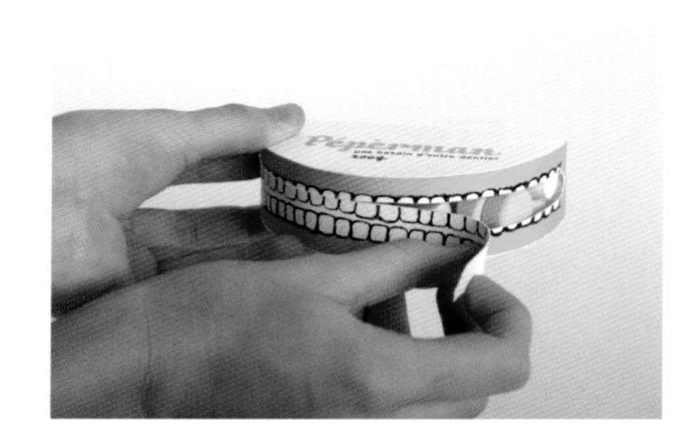

studio **Atelier BangBang**
Montréal Canada
designer Simon Laliberté
www.atelierbangbang.ca

This product was made in a packaging class and for the contest Young Package 2011, whose theme was: packaging of a national product. The peppermint reminds us of our grandparents who ate these infamous candies without the help of their dentures. This is where the idea comes from - unwrapping the product by tearing the teeth apart. This humoristic packaging requires very little glue, cardboard and a small recyclable plastic wrap.

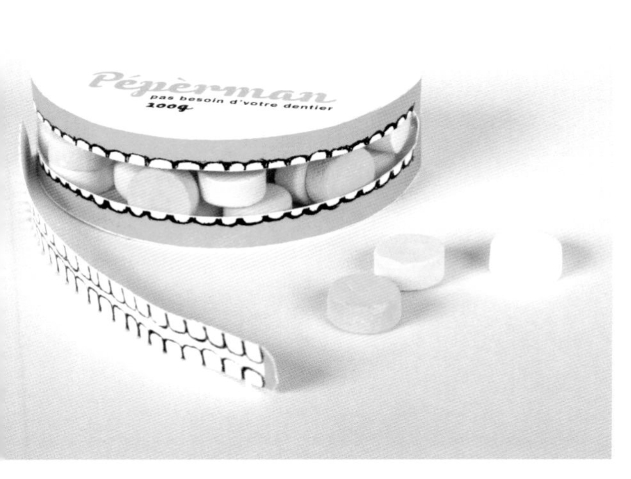

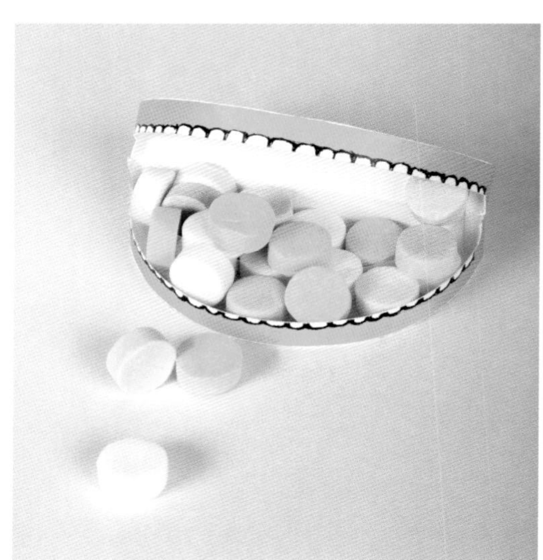

Herokid™ Magic Box

studio **Andreu Zaragoza**
Barcelona Spain
designer Andreu Zaragoza
illustrations Registred Kid, Aryz & Grito
www.andreu-zaragoza.com

Package designed for the urban clothing brand Herokid.

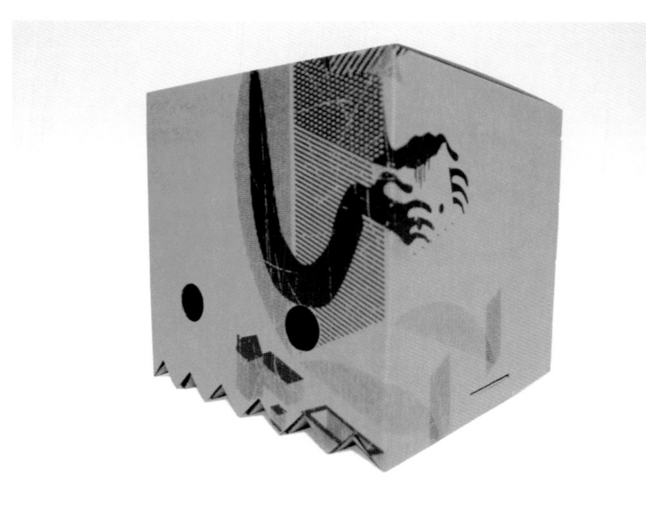

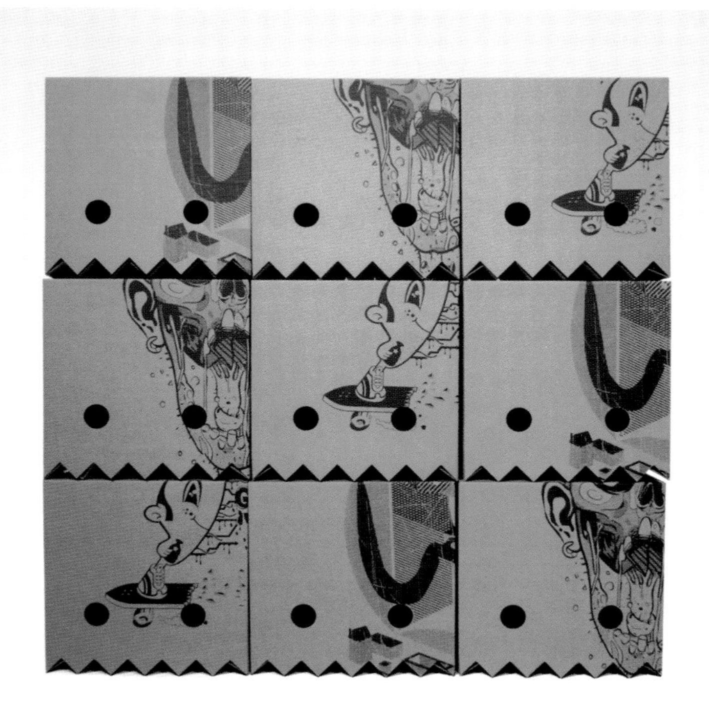

Morfoze Polyhedron

MORFOZE

agency **Yevgeny Razumov**
Moscow Russia
designer Yevgeny Razumov
www.behance.net/razumov

MORFOZE is a hard soap concept, which will surely be liked by all who have had to face the principles of 3D-modelling: software designers, modellers, engineers and people who like everything unusual and anything that makes our lives more interesting. This polygonal soap is odor-free, like everything that exists in the virtual space, and it illustrates the form transformation. Its smoothness refers to the processes of creating and editing objects in three-dimensional programs. Even with all its 'game' it doesn't lose its main function - to clean and moisten the skin.

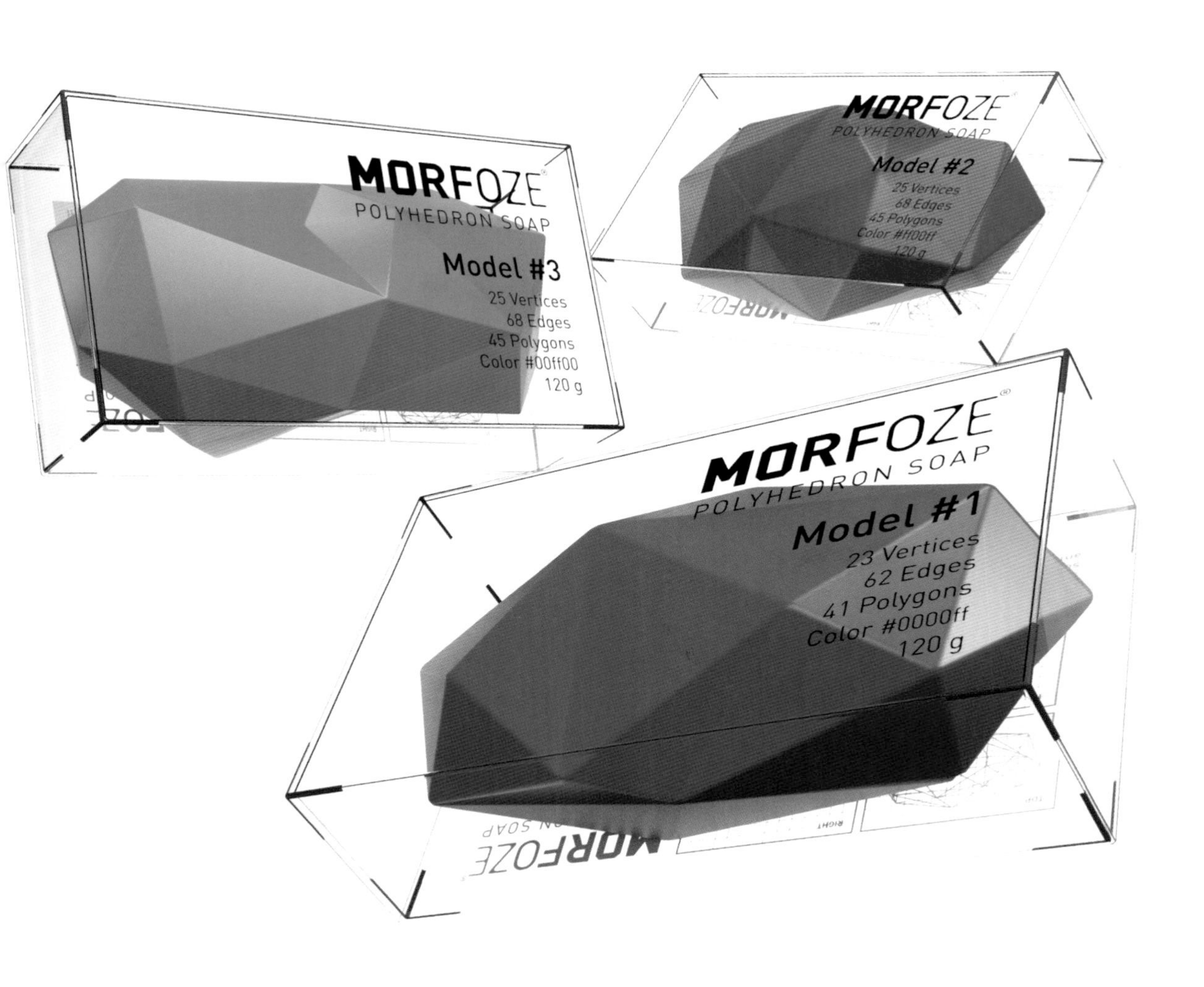
MORFOZE
POLYHEDRON SOAP
Model #3
25 Vertices
68 Edges
45 Polygons
Color #00ff00
120 g
MORFOZE
POLYHEDRON SOAP
Model #2
25 Vertices
68 Edges
45 Polygons
Color #ff00ff
120 g
MORFOZE
POLYHEDRON SOAP
Model #1
23 Vertices
62 Edges
41 Polygons
Color #0000ff
120 g

Fedrigoni gemstones

studio **Lun Yau**
London UK
designer Lun Yau
www.lunyau.co.uk

The concept was to produce a fully integrated campaign to launch and promote Fedrigoni's "Imaginative Colors" paper selection tool in the UK. To create a campaign that appeals to the design industry by creating a spectacle of the colored paper. The paper was displayed as not simply a flat product, but as a creative, tangible and versatile medium. Members of the industry were invited to discover the colors and precious papers of Fedrigoni's "Imaginative Colors" paper selection tool.
The colors and paper were presented by exhibiting precious gemstones that are discovered inside the box. The discovery promotes the color selection tools and shows the quality and capablities of Fedrigoni's papers to the design industry. Representing each of the four "Imaginative Colors" tools are four gemstones crafted from vivid colors. The facets of the gemstones contain different colors of paper that make the gemstones appear shiny and glimmery.

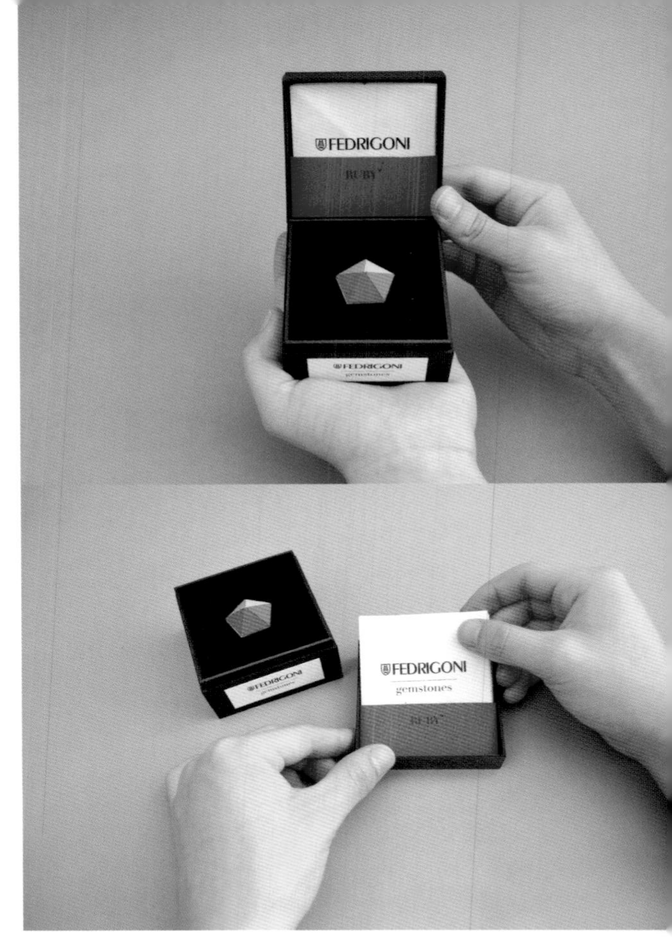

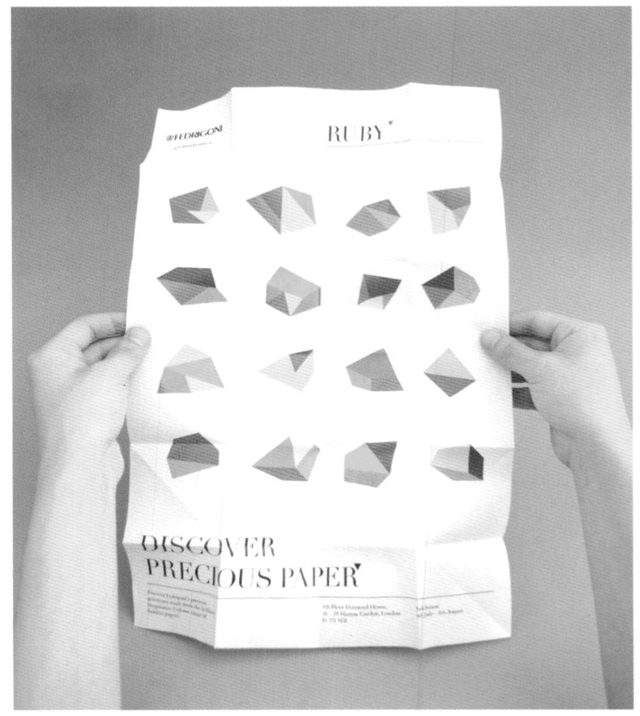

FEDRIGONI
DIAMOND
FEDRIGONI
RUBY
FEDRIGONI
TOPAZ
FEDRIGONI
SAPPHIRE

FEDRIGONI
gemstones

Take Away

designer Carolina Caycedo V.
Cali Colombia
cayca15@hotmail.com
designer Marisol Escorza H.
Chile Spain
www.mescdesign.com

The purpose of this project was to create a packaging that sold gourmet croquettes. From the formal point of view, the packaging needs to be multifunctional, capable of holding, transporting and serving in different situations (walking, reading a book, to take away, etc.).
Graphically, the title evokes a reality, an analogy that incorporates photographs that take us to a certain point in history, accompanied by a text praising the croquette as such and giving it a domestic touch through the use of the napkin.

Menuda, humilde, dorada, crujiente, aromática, sabrosa.
La croqueta es un acierto, un logro, una genialidad.
Un bocado del Paraíso empanado y frito. Quien prueba una
buena croqueta no puede resistirse al hechizo
de coger otra, y otra, y otra...
No Cocino Más!
TAKE AWAY

Menuda, humilde, dorada, crujiente, aromática, sabrosa,
La croqueta es un acierto, un logro, una genialidad.
Un bocado del Paraíso empanado y frito. Quien prueba una
buena croqueta no puede resistirse al hechizo
de coger otra, y otra, y otra...
No Cocino Más!
TAKE AWAY

Menuda, humilde, dorada, crujiente, aromática, sabrosa,
La croqueta es un acierto, un logro, una genialidad.
Un bocado del Paraíso empanado y frito. Quien prueba una
buena croqueta no puede resistirse al hechizo
de coger otra, y otra, y otra...
No Cocino Más!
TAKE AWAY

Sweet&Hot

studio **The British Higher School of Art and Design**
Moscow Russia
con?ept Shashkina Ivanna
design & lettering Shashkina Ivanna
ivanna.shashkina@gmail.com

This is a project by a student from The British Higher School of Art and Design: a packaging of chocolate with spices.
Traditionally, chocolate packaging is very formal, even for chocolate with spices. But I wanted to reflect the 'sharp' and 'unusual' nature of these spicy chocolates. At first I tried using sharp creases to mirror the sharpness of the taste, but that felt rigid and boring. It lacked emotion.
So I took an alternative approach instead: using the involuntary and humorous faces that people make when tasting spicy foods.
The brand-name "Sweet & Hot" reflects the nature of the product: it's an assortment of chocolates with a variety of hot spices. These chocolates are marketed for young people and the packaging depicts young women's faces reflecting the 'heat-levels' of the chocolate. Each paper bag contains 20 pieces of heart-melting dark chocolate. Advertising posters for this product are designed to be funny and provocative - aimed at young people in their early twenties who enjoy freedom and new experiences.

YOU CAN SHARE WITH EVERYONE
20 PIECES
PIMENTA
CHOCOLATE WITH SPICES

BE CAREFUL!
20 PIECES
HABANERO
CHOCOLATE WITH SPICES

The real cookbook

agency **Kolle Rebbe / KOREFE**
Hamburg Germany
creative director Antje Hedde
designers Reginald Wagner, Christine Knies
idea Gereon Klug
www.korefe.de

The Real Cookbook is the first cookbook that really lives up to its name: it's a book you can actually cook and eat. Every page is made of fresh pasta. With a little sauce between the pages and a sprinkle of cheese on top, it becomes lasagne. The Real Cookbook is a special edition from the Gerstenberg Publishing House, specialists of high-quality culinary and art books. The prose etched on the four inner pages of pasta, toys with the idea of how important the contents of a cooking book can be.

echte
UND
EINZIGE
KOCHBUCH
GERSTENBERG
DAS
echte
UND

echte
UND
EINZIGE
KOCHBUCH

RUBBER GLOVES
wtp
PAINT ROLLER
wtp
PAINT ROLLER
wtp
PROFESSIONAL
GUMBOOTS
GUM
BOILERSUIT
BOILERSUIT

MR. GOO
SUPREME GREEN
100% BIODEGRADABLE
MR. GOO
100% BIODEGRADABLE
MR. GOO
EXTREME ORANGE
100% BIODEGRADABLE
MR. GOO
BLAZING RED
100% BIODEGRADABLE
MR. GOO
PUNCHED-OUT PURPLE
NET WT: 120g
100% BIODEGRADABLE
MR. GOO
CHILLING BLUE
NET WT: 120g
100% BIODEGRADABLE

Packages with unique illustrations

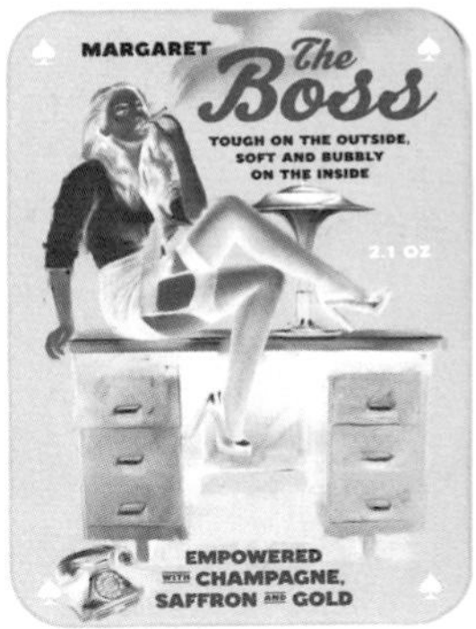

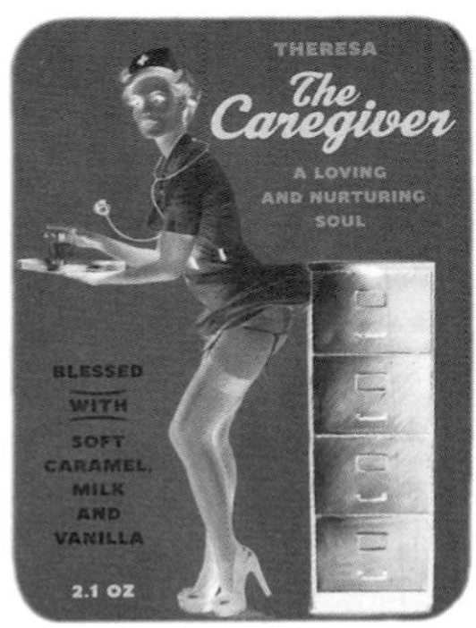

studio **Julia Castaño**
Barcelona Spain
designer Julia Castaño
www.juliacastano.com

Magnum is a brand of ice-cream bars that are covered by a thick layer of chocolate. It is a product aimed at the adult public, but the company wanted to focus on the younger public with a new line of ice-cream bars. Magnum's well-known slogan is "intense pleasure." Now, the same type of slogan is used again, but applied to the young ones. For a child there is nothing that gives more intense pleasure than having fun. So the new line Donec was born, full of family characters made of milk and fruit, covered in a thick layer of chocolate.
Donec is the new generation of Magnum (from the latin "Great"), and also based on latin meaning "fun." The ice-cream comes in a cardboard box with a die-cut drawing of a family character, which allows the child to cut it and place it on the ice-cream bar stick, to use as a puppet once he or she is finished eating the ice-cream.
Inside the box with the ice-cream, the child can find eatable accessories that belong to the ice-cream character. By this, the child is interacting with the product, making him or her create the character for themselves.

Félix
DONEC
DONEC

Félix
DONEC
DONEC

DONEC

Félix
DONEC
MAGNUM

DONEC

Paula
DONEC
MAGNUM

Martín
DONEC
MAGNUM

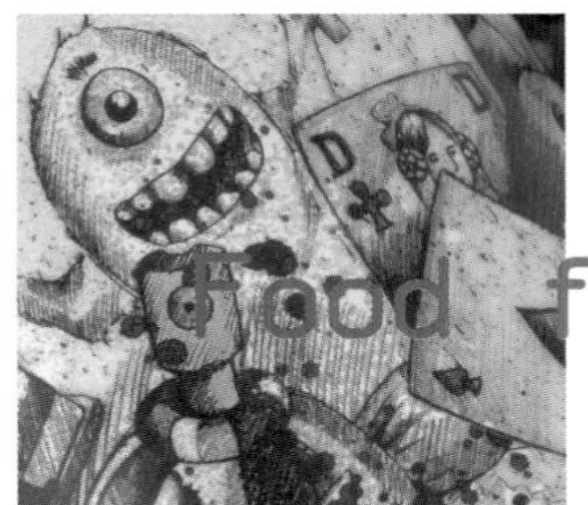

Food finish

agency **Kolle Rebbe / KOREFE**
Hamburg Germany
creative director Katrin Oeding
designer / illustrator Reginald Wagner
photography Ulrike Kirmse
www.korefe.de

Food Finish is the world's first food coloring in a spray can. Available in four colors – gold, silver, red and blue – it can be used to refine all manner of luxurious feasts.
The Deli Garage is a food label that sells delicatessen products with the design and practicality of everyday DIY and garage items.

THE DELI GARAGE
ESSLACKIEREREI
VIER LEBENSMITTELSPRÜHFARBEN
THE DELI GARAGE
FOOD COOPERATIVE
ESS
LACK
GOLD
SILBER
BLAU
ROT

THE DELI GARAGE

Bastoncillos

Student Work / HealthCare

designer Federico Beyer
Medellín Colombia
www.behance.net/packfedericobeyer
designer Marisol Escorza H.
Chile Spain
www.mescdesign.com

This project proposes to rethink the design for Johnson & Johnson's Q-tip container. The design has been renewed in terms of industrial design, safety and accessibility. This amusing concept aims to distract the child's attention and to invite her to participate in the cleaning process - perhaps through telling a story based on the featured character. The illustrations are inspired by the work of California based illustrator, Hsinping Pan.

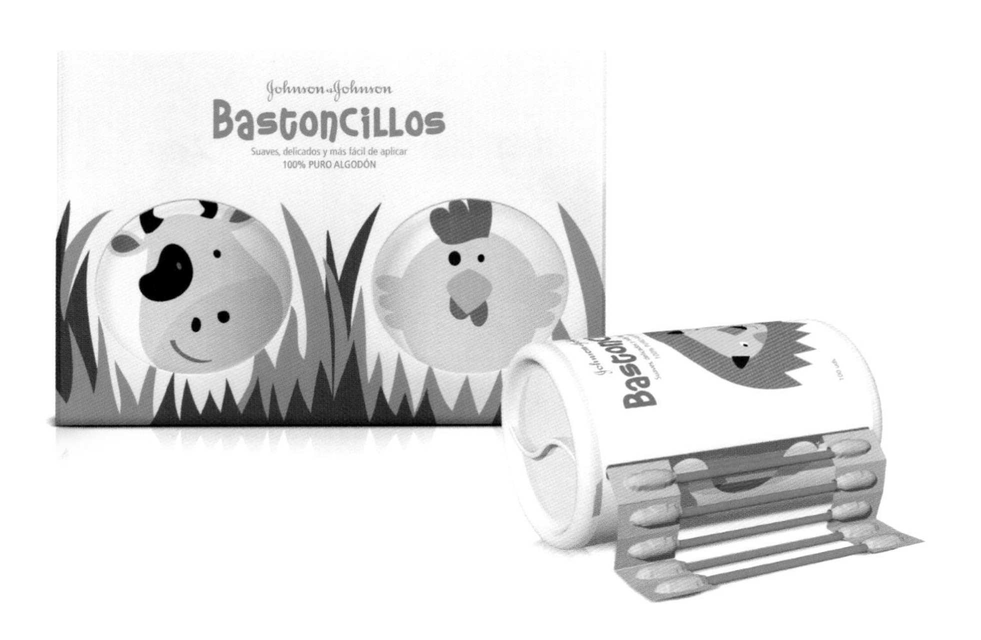
Johnson & Johnson
Bastoncillos
Suaves, delicados y más fácil de aplicar
100% PURO ALGODÓN

Johnson & Johnson
Bastoncillos
Suaves, delicados y más fácil de aplicar
100% PURO ALGODÓN
100 uds.

vamos a la granja al atardecer
curiosos animales vamos a conocer
muuuu!

Johnson & Johnson
Bastoncillos
Suaves, delicados y más fácil de aplicar
100% PURO ALGODÓN
100 uds.

vamos a la selva al amanecer
curiosos animales vamos a conocer
UH UH UH!

Johnson & Johnson
Bastoncillos
Suaves, delicados y más fácil de aplicar
100% PURO ALGODÓN
100 uds.

vamos a la selva al amanecer
curiosos animales vamos a conocer
Grrr!

Johnson & Johnson
Bastoncillos
Suaves, delicados y más fácil de aplicar
100% PURO ALGODÓN
100 uds.

vamos a la granja al atardecer
curiosos animales vamos a conocer
Cocoro coo!

Haircare Xpressions

agency **Gworkshop Design**
Quito Ecuador
designer José Luis García Eguiguren
www.gworkshopdesign.com

The purpose of this project was to utilize the container shapes of the brand HAIRCARE to create a new product line for young adults. For this, the theme "Xpressions" is aimed at inviting the consumer to experience something new. The typography interacts with the package, giving it a modern and unique style that ties the product line together. A subtle alteration in the type, that changes according to the composition of the phrases, creates a major visual impact and helps associate with the personality of the brand. The shapes and images form hair styles and mix with the facial expressions of people, to depict the products' purpose. The idea is to print these images on matte metallic containers.

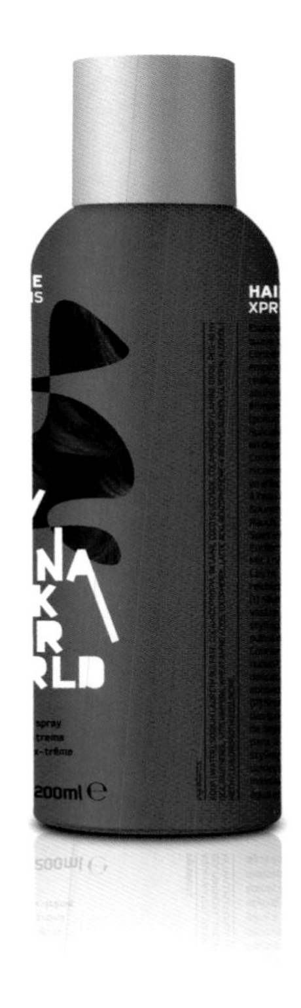

HAIR.CARE
XPRESSIONS
I WILL BLOW YOU UP
HARD GLUE
Rock Spiking Gel
Gel efecto rock
Gel effet rock
1.7 fl.oz. | 50ml ℮

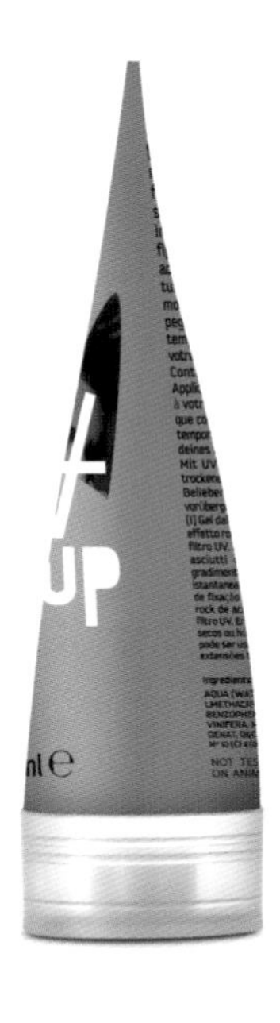

Ingredients:
NOT TESTED ON ANIMALS

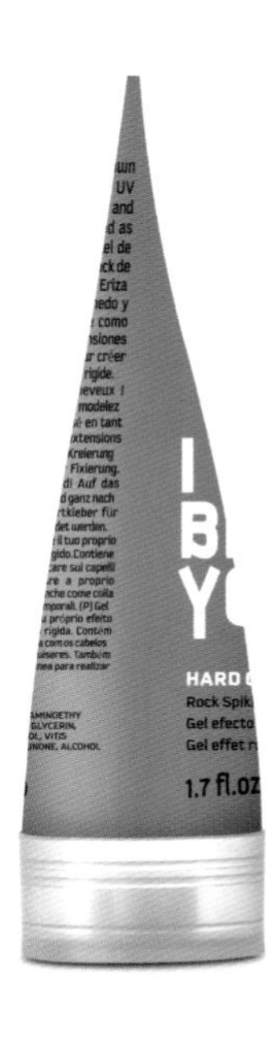

I DON'T SHINE, BUT YOU WILL
HAIR.CARE
XPRESSIONS
CLAY WAX

I AM COLD, BUT YOU ARE HOT
HAIR.CARE
XPRESSIONS
SCULPTING ICE

IT'S ALL ABOUT SEXY CURVES
HAIR.CARE
XPRESSIONS
CURLY TOUCH

HAIR.CARE
XPRESSIONS
I WILL MAKE YOU GO CRAZY
STOP FRIZZ

I DON'T SHINE, BUT YOU WILL

I AM COLD, BUT YOU ARE HOT

IT'S ALL ABOUT SEXY CURVES

I WILL MAKE YOU GO CRAZY

Chocolates with Attitude vol·II

studio **Bessermachen Designstudio**
Copenhagen Denmark
creative director Kristin Brandt
illustrator Niels Ditlev
www.bessermachen.com

For the second year in a row, Brandhouse and Bessermachen Design Studio have developed a box with 12 different chocolates named "Chocolates with Attitude." The chocolate is created as a manifesto about brand personalities and packaging design.
The challenge was to illustrate brand personalities and archetypes in an innovative way. The creative solution was a 3.5 kilo chocolate case with 12 different tin-boxes, each representing an archetype or personality in the form of a pin-up girl. Thereby each delicious chocolate reflects an individual archetype in both the ingredient composition and execution. In addition, a 2012 calendar was included, with the 12 pin-up girls each representing a month
Every flavor composition was designed for an archetype and handcrafted by Henrik Konnerup from Konnerup & Co. The hero, as a pinup girl, was illustrated as a firefighter, containing "ganache with black pepper, balsamic vinegar and covered with 70% dark chocolate."

Chocolates
with
attitude
BY BRANDHOUSE GROUP

Chocolates
with
attitude
The Girl next door
The Innocent
The Entertainer
The Creator
The Boss
The Hero
The Adventurer
The Rebel
The Professor
The Caregiver
The Magic lady
The Seducer

ANNA
The Girl next door
WHAT YOU SEE IS WHAT YOU GET
THE ORIGINAL TASTE OF REAL MARZIPAN AND DARK CHOCOLATE
2.1 OZ
MARY
The Innocent
ALL NATURAL, ALL PURE AND ALL GOOD FOR YOU
COMES WITH SPRING WATER, LAVENDER AND ORGANIC SILK WHITE CHOCOLATE
2.1 OZ
2.1 OZ
BILLIE
The Entertainer
ALWAYS TUNED IN AND TURNED ON
CREAMY AND LIGHTLY SALTED WITH PURE LIQUORICE AND MILK CHOCOLATE
ROSE
The Creator
A CRUNCHY, INSPIRING MOUTHFUL
CRUSHED ALMONDS WITH A FINE MIX OF CHOCOLATES
2.1 OZ
MARGARET
The Boss
TOUGH ON THE OUTSIDE, SOFT AND BUBBLY ON THE INSIDE
2.1 OZ
EMPOWERED WITH CHAMPAGNE, SAFFRON AND GOLD
LAURA
The Hero
FLAVOURED WITH ASHES, BLACK PEPPER AND BALSAMICO
2.1 OZ
SALLY
The Adventurer
A TASTY TRIP AND AN EXHILARATING EXPERIENCE
REFRESHED WITH PEPPERMINT AND EXOTIC PISTACHIO
2.1 OZ
AMY
The Rebel
DRY, SMOKY AND FULL-BODIED
WITH A DROP OF SINGLE MALT WHISKY
CALIFORNIA
US
66
2.1 OZ
MISS JONES
The Professor
TO BE OBEYED AND ENJOYED
RESPECTFULLY FILLED WITH SUPER FRUIT, DRIED SEAWEED AND GINGER
2.1 OZ
THERESA
The Caregiver
A LOVING AND NURTURING SOUL
BLESSED WITH SOFT CARAMEL, MILK AND VANILLA
2.1 OZ
TRIXIE
The Magic lady
FILLS YOU WITH ILLUSIONS AND CHARM
PUFFING, BLUE CHOCOLATE WITH ALMOND NOUGAT
2.1 OZ
BELLA
The Seducer
TEMPTING WITH A NATURAL SWEETNESS
RASPBERRY AND ROSE WATER ADDED WITH PASSION
2.1 OZ

Chocolates with Attitude vol.II

studio **Bessermachen Designstudio**
Copenhagen Denmark
creative director Kristin Brandt
illustrator Niels Ditlev
www.bessermachen.com

Calendar with attitude
AND THE STORY BEHIND
BY BRANDHOUSE GROUP
THE UPGRADING COMPANY

BILLIE
The Entertainer
November
2012
1 2 3
4 5 6 7 8 9 10
11 12 13 14 15 16 17
18 19 20 21 22 23 24
25 26 27 28 29 30
Calendar with attitude

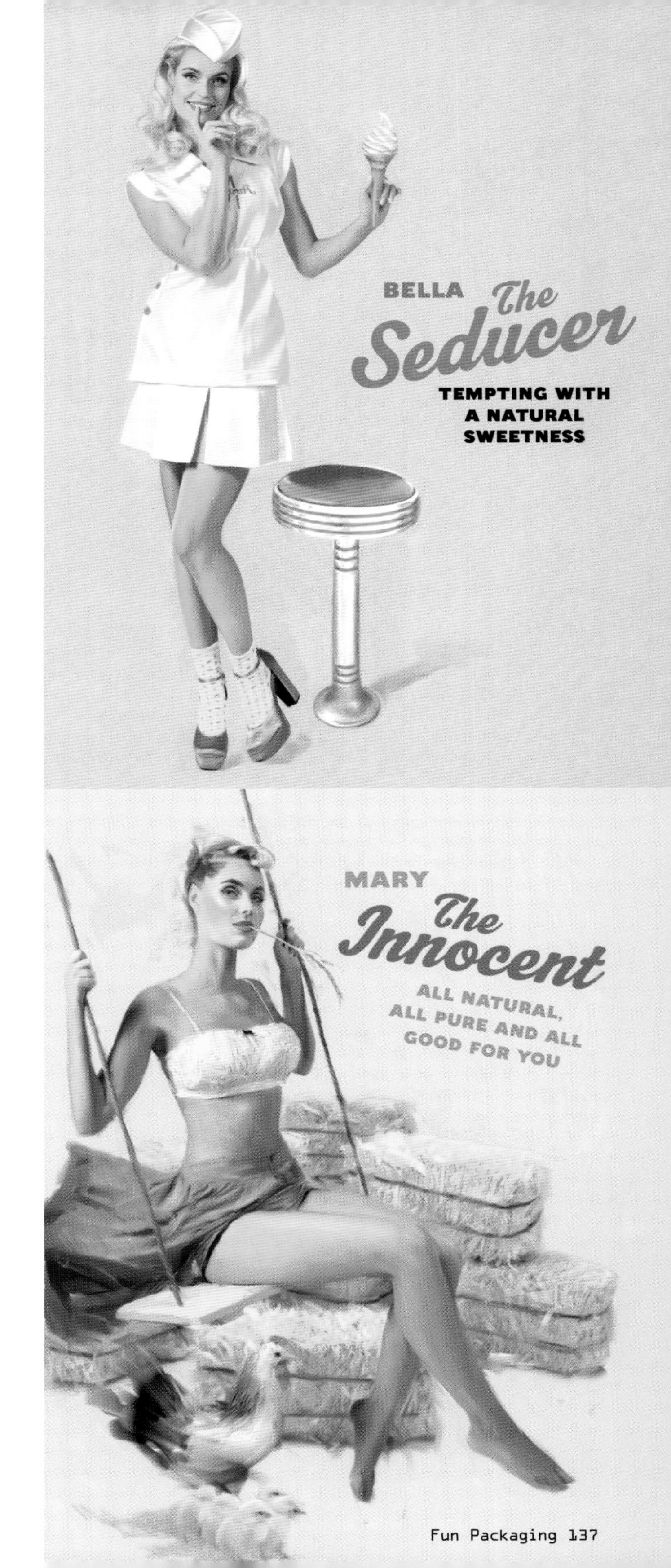
BELLA
The Seducer
TEMPTING WITH
A NATURAL
SWEETNESS
MARY
The Innocent
ALL NATURAL,
ALL PURE AND ALL
GOOD FOR YOU

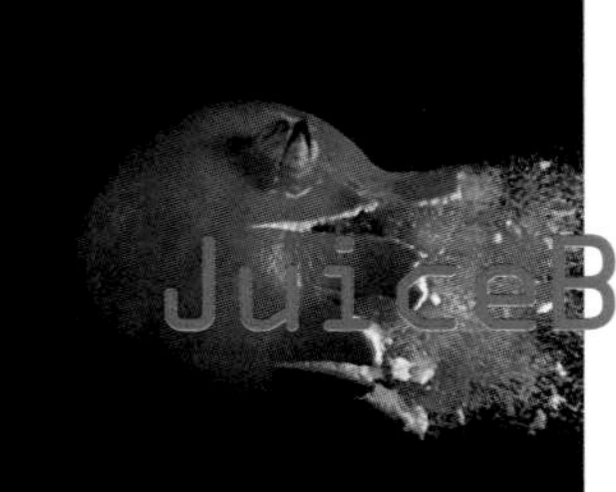

JuiceBurst

agency **Williams Murray Hamm**
London United Kingdom
creative director Garrick Hamm
design director Grant Willis
designer Grant Willis & Rachel Price
account director Wybe Magermans
director of photography Matt Broad
special effects (pyrotechnics) Artem
www.williamsmurrayhamm.com

Step forward, JuiceBurst, a brilliant, inexpensive product in big, glugging sizes with a wide range of flavors. Aimed at 18 -24 year olds, it has to be a beacon of honest goodness amongst dusty second tier brands.
The design uses a series of copy 'outbursts' such as, "any more goodness, I'd be a nun!" set against images of fruit bursting on the pack. The idea is in the name.
Work included creating new pack structures and producing films of the bursting fruit that play a key role in the interactive labelling that, via an app, offers fruit machine games or a short film of how we detonated each fruit.

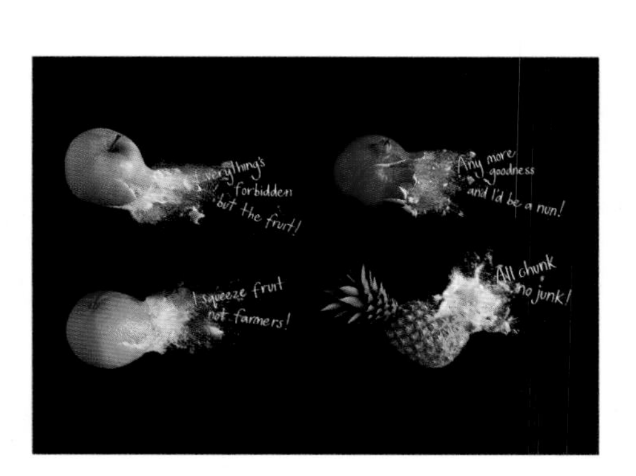

JUICEBURST
MANGO
JUICEBURST
BLOOD ORANGE
JUICEBURST
JUICEBURST
Anymore goodness
and I'd

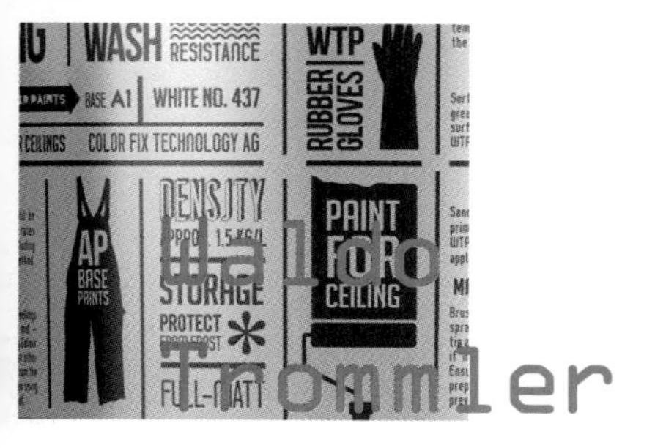

Waldo Trommler

agency **Reynolds & Reyner**
Kiev Ukraine
creative directors Artyom Kulik,
Alexander Andreyev
www.reynoldsandreyner.com

How to create a brand that stands out? We need to find a design solution that hasn't been used by any of the competitors, while at the same time showing the main features of the company – friendliness, quality and innovation. WTP is not just a manufacturer of paints – it's an assistant, always ready to help, suggest and defend from hassles and problems. Repairs with WTP are simple, convenient and fast. This shows in its simple design.

No doubt, WTP is the most friendly and remarkable brand of paints on the shelf now. Every item is bright and a memorable combination of colors and objects, which together form the entire brand.

wtp
ACADE

wtp
SUPERIOR HEAT RESISTANT
ENAMEL FOR RADIATORS
9L
ADIATO
SPECIAL ALKYD PAINT FOR INTERIOR
AND EXTERIOR METAL SURFACES
METAL
PAINT FOR INTERIOR CONCRETE
AND WOODEN FLOORS
FLOOR
WATERBORNE ACRYLATE
LATEX PAINT
WALL
THRO
LING

Waldo Trommler Paints

agency **Reynolds & Reyner**
Kiev Ukraine
creative directors Artyom Kulik, Alexander Andreyev
www.reynoldsandreyner.com

wtp
WATERBORNE ACRYLATE
LATEX PAINT
9L
CEILING

CEILING
WALDO TROMMLER PAINTS
WASH
RESISTANCE
BASE A1
WHITE NO. 437
COLOR FIX TECHNOLOGY AG
WORK WITH
WTP
RUBBER GLOVES
APPLICATION CONDITIONS
PREPARATION
PAINTING
MAINTENANCE INSTRUCTIONS
CLEANING OF TOOLS
MATE
SINCE 1966
PAINT FOR WALLS
BIO PROTECT
AP
DENSITY
APPROX. 1.5 KG/L
STORAGE
PROTECT
FULL-MATT
PAINT FOR CEILING
CEILING
CAN SIZES

wtp
WOOD
wtp
9L
FLOOR
wtp
WATERBORNE ACRYLATE
LATEX PAINT
9L
WALL

Brio & Vardo detergent

agency **Studio h**
London United Kingdom
design & illustration Rob Hall
www.studioh.co.uk

Brio and Vardo, two new brands that are the first detergent products to be manufactured in Georgia. Aimed initially at the Georgian market and their neighboring countries, it was important that Brio and Vardo appealed as local brands reflecting their Georgian provenance, while at the same time competing on shelf with the large global competitors. Standing aside from the industry norms, Studio h used symbolic imagery to create an emotional link to the culture. The rose, Georgia's national flower, was developed as a strong, fresh identity for Vardo and the circular Brio illustration portrays the Georgian landscape of mountains and trees with a color palette referencing the natural hues of the Southern Black Sea region.

ავტომატი უნივერსალური მთის სურნელით
ვარდო
450 გრ
Berla

უნივერსალური ხელით რეცხვისთვის
ვარდო
მთის სურნელით

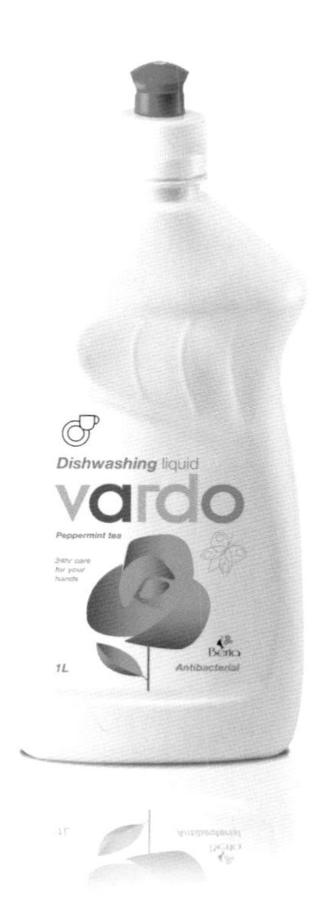
Dishwashing liquid
vardo
Peppermint tea
1L
Berla
Antibacterial

universal automatic
mountain freshness
vardo
450g
Berla

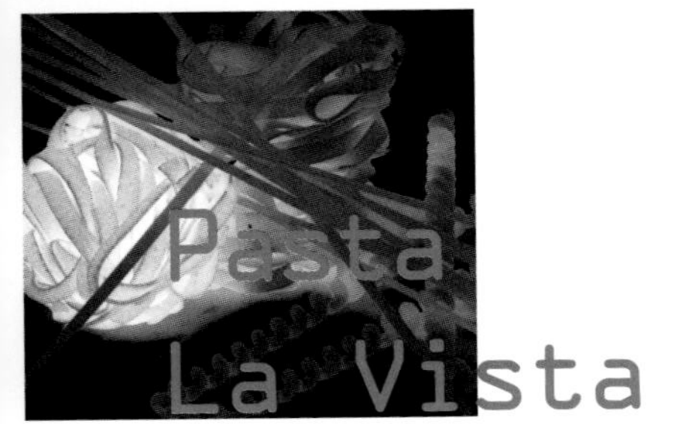

Pasta La Vista

studio **Andrew Gorkovenko**
Moscow Russia
designer Andrew Gorkovenko
www.gorkovenko.ru

Pasta La Vista – a brand that covers a wide range of various hand-made macaroni products manufactured in accordance with traditional Italian recipes and using only ecologically clean products of the highest quality.
Considering the key feature of the brand, which is hand-made manufacturing, we decided to introduce four characters into the corporate style and package. The pasta is made by particular people, they are: Mario, Francesco, Giovanni and Francesca. Therefore, every package depicts one of the characters – an Italian chef in the process of cooking.
In addition to that, the product in the package appears to be the hair of the characters, neatly tucked under the cook's cap, which also serves as a package element that provides access to the product. All of this allows the consumer to receive pasta directly from the 'hands' of the Italian chef.

PASTA ITALIANA
Pasta
La Vista
PASTA ITALIANA
Pasta
La Vista
PASTA ITALIANA
Pasta
La Vista

Olive oil

agency **Mousegraphics**
Athens Greece
creative director Gregory Tsaknakis
designer Vassiliki Argyropoulou
www.mousegraphics.gr

Our client requested that we design a packaging that would communicate the premium quality of the product and differentiate it from its competition.
This limited edition premium olive oil employs a tin-can rather than the conventional glass bottle. Its packaging deliberately moving away from traditional symbols of olive oil quality or clichés of provenance. The end product targets the mind; aesthetics is the real reason to buy. Almost as an afterthought, a very realistic-looking drop of oil on the tin is what keeps it grounded in the food section of a super market. If ever there was a slogan attached to it, then this would read "simply olive oil."

Oil
WITH TRUFFLE

Oil
WITH ORANGE

Oil
WITH BASIL

ILIADA EXTRA
VIRGIN OLIVE
Oil
TRUFFLE FLAVOUR
O!
LIVE
Oil

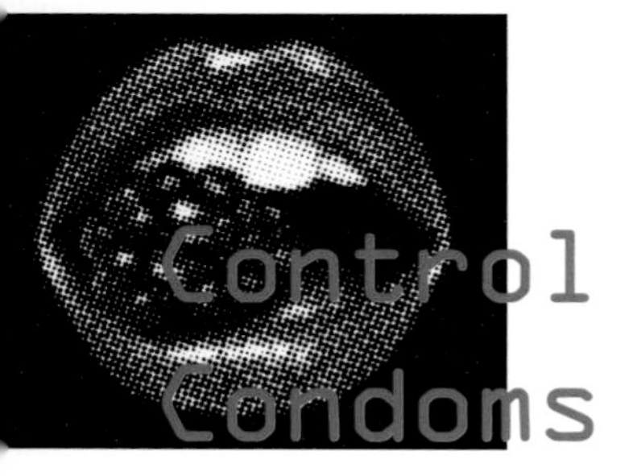

Control Condoms

studio **Student Work**
Barcelona Spain
designers Mara Rodríguez & Catarina Pinheiro
www.oloramara.com
www.behance.net/CatarinaPinheiro

The purpose of this project was to redesign Control Condoms packaging and branding.
The packaging gives you the idea of a pill, so you have the sense of more security, which gives you more confidence in the product. There are two packs, one bigger to keep at home and another one with four condoms to take away.
About the graphics: we chose the idea of lips to represent sensuality and sexuality. We gave different lips and typography colors to each product and we composed it over a black background, giving a smart feeling to the packs. The black background also helps people think about nightime and adult pleasures.

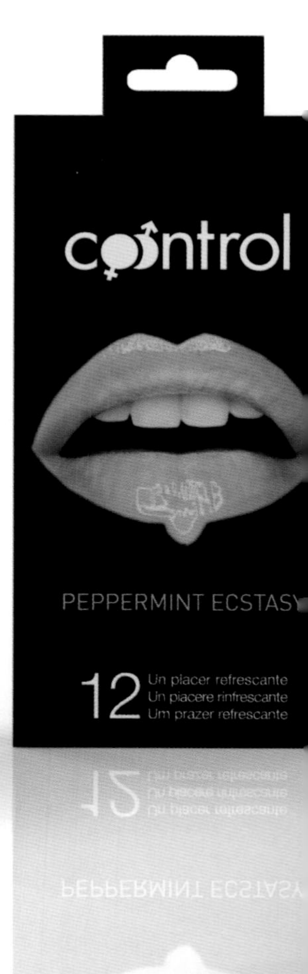

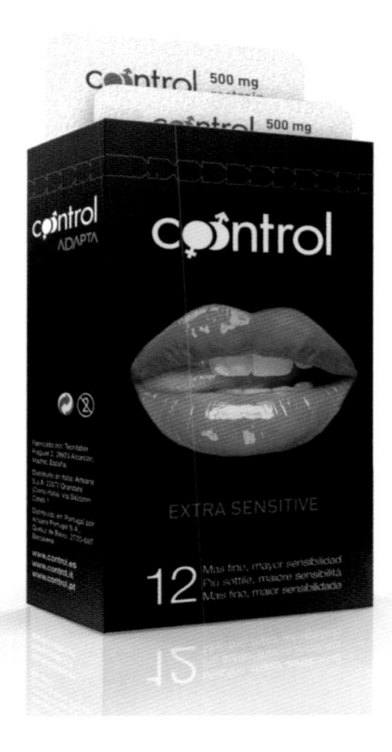
control
ADAPTA
control
EXTRA SENSITIVE
12 Más fino, mayor sensibilidad
Più sottile, maiore sensibilità
Mais fino, maior sensibilidade
www.control.es
www.control.it
www.control.pt

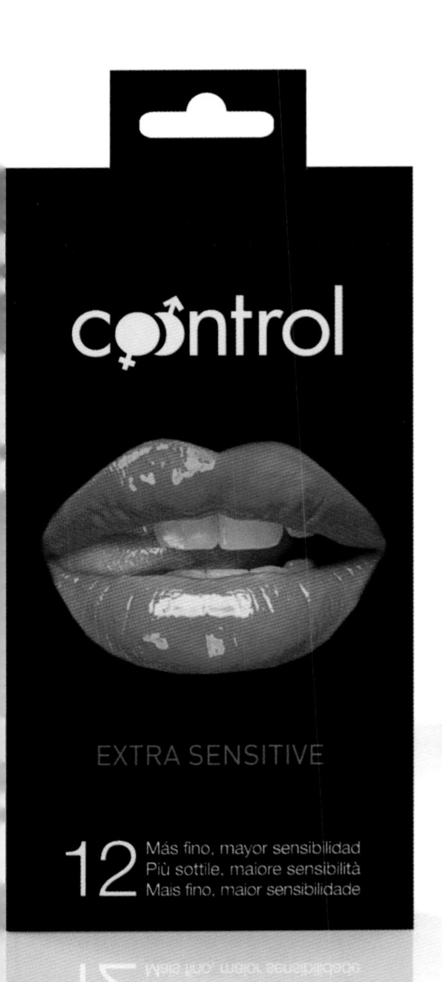
control
EXTRA SENSITIVE
12 Más fino, mayor sensibilidad
Più sottile, maiore sensibilità
Mais fino, maior sensibilidade

control
INTENSE FRUIT
12 Un dulce placer afrodisiaco
Un dolce piacere afrodisiaco
Um doze pracer afrodisiaco

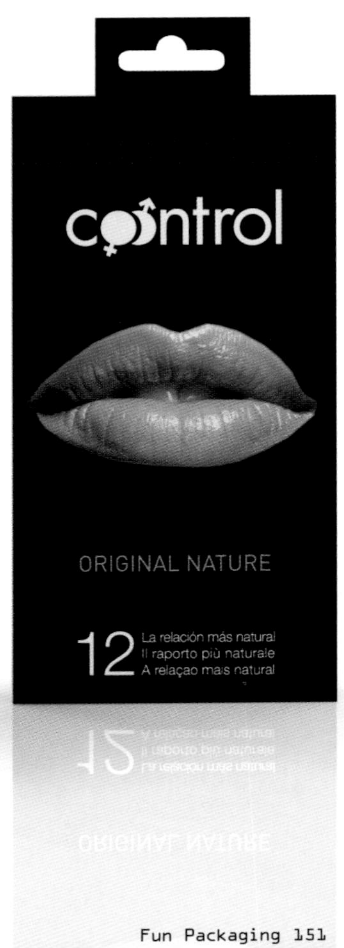
control
ORIGINAL NATURE
12 La relación más natural
Il raporto più naturale
A relaçao mais natural

Fruit Fusion

agency **Lavernia & Cienfuegos**
Valencia Spain
www.lavernia-cienfuegos.com

The active ingredients of fruit have nutritive properties and vitamins that are good for caring for the skin. The main objective made by the brief was to maximize the presence of fruit and to add new sensations and experiences to the routine of skin care and beauty. The idea was to develop the design based on the concept of gourmand, or overindulgence. A juicy and inviting cocktail of fruits are able to bring together ideas such as, nutrition, health, natural, tasty, aromatic...and translate them into the field of cosmetics. The design should be clear, straightforward and very effective.

Moscow Russia
designers Galima Akhmetzyanova & Pavla Chuykina
3D visualisation Vladimir Pospelov
www.cargocollective.com/hellogalima
www.behance.net/pavla

The brand name comes from an ancient female name, which literally means domination. Russian women bear the double burden of a job and family-raising responsibilities, in which husbands generally participate little. So, there are three stories about a woman and her daily routine.
We admire traditional art and believe that it has a lot of prospectives for graphic designers and artists. Russian embroidery is by far the most interesting field because the images seem very archaic but, at the same time, are very contemporary. In our project we used traditional elements such as grass and birds and colored them differently from traditional red and yellow.

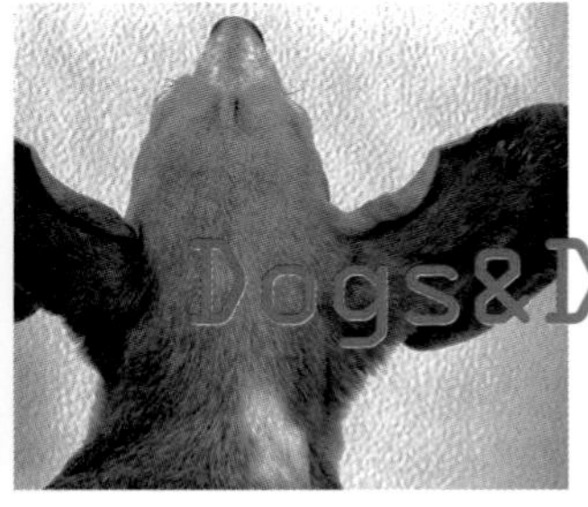

Dogs&Drops

studio **Student Work**
Barcelona Spain
designers Mara Rodríguez & Catarina Pinheiro
www.oloramara.com
www.behance.net/CatarinaPinheiro

Dogs&Drops is a dog care brand that takes care of our pets and keeps them clean and happy. We represented the bath moment with a funny element that will attract dog owners. For that, we have used dog photographs where dogs are playing with different bath elements. The bottle appears clean and vibrant. Attractive colors were added to make the brand look young and fresh.

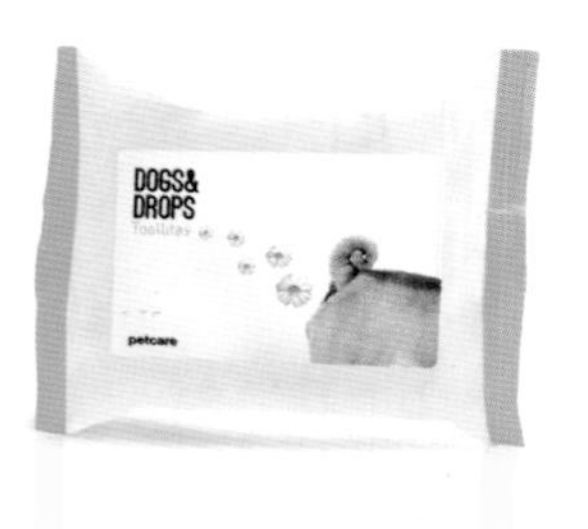

DOGS&
DROPS
Long hair
Shampoo
Pelo largo
Champú
10.2 fl.oz./300 ml

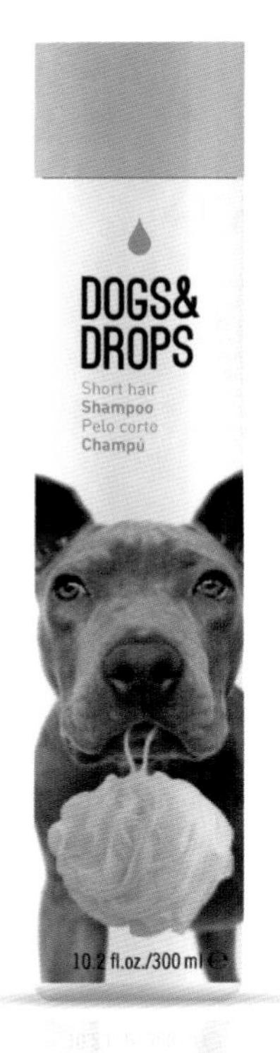
DOGS&
DROPS
Short hair
Shampoo
Pelo corto
Champú
10.2 fl.oz./300 ml

DOGS&
DROPS
White hair
Shampoo
Pelo blanco
Champú
10.2 fl.oz./300 ml

DOGS&
DROPS
It will love it!
petcare
DOGS&
DROPS
Tarjeta Regalo
petcare
DOGS&
DROPS
It will love it!

Jade Monk

agency **Moxie Sozo**
Boulder, CO USA
designer Charles Bloom
www.moxiesozo.com

An Austin, Texas-based startup with big dreams of creating a profitable, successful business from selling matcha green tea beverages, contacted Moxie Sozo to create their company. From the naming, to the package and web designing, Moxie Sozo took the lead on crafting Jade Monk into the brand it is today. The guys behind Jade Monk wanted an edgy, modern take on an ancient Japanese beverage. The overall look and feel – replete with electric hues of pinks, greens and oranges – alluded to the cultural significance, while downplaying the more traditional elements. With that in mind, Moxie Sozo designed four tins to house four flavors of the tea that featured individually inspired Japanese folkloric creatures. Moxie Sozo also produced the Jade Monk website to accurately and quirkily convey the brand's wonderful tale. The result is a strong and feisty brand that's found its niche in the world as a mighty tasty brew.

Monk
PALAU

FLAVORED MATCHA TEA
CHAI SPICE
Jade Monk
STONE GROUND
MATCHA
GREEN TEA POWDER
CHAI SPICE

Jade Monk
MATCHA TEA
LIME
BLOSSOM
MATCHA TEA

Jade Monk
MATCHA TEA
PALAU
PEACH
MATCHA TEA

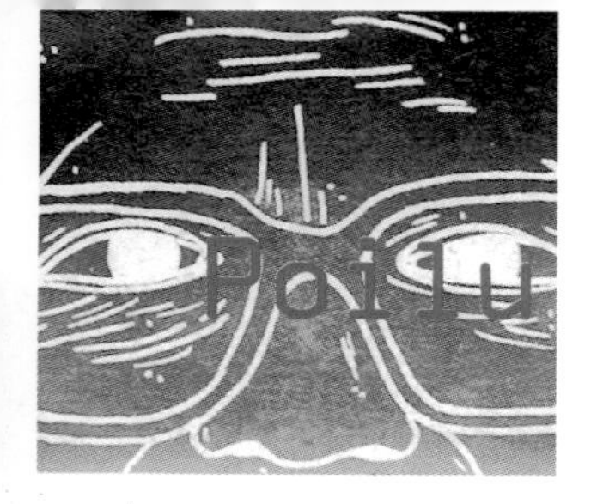

studio **Atelier BangBang**
Montréal Canada
designer Simon Laliberté
www.atelierbangbang.ca

This humoristic packaging offers the function of assembling two products (two paintbrushes) together with only one cardboard printed on both sides. One paintbrush is a big one and the other is a small one, for finishing touches.
Once it's folded, the package has two utilities: One, protecting paintbrushes when they are shipped; and two, supporting the paintbrushes when they are full of paint. Each paintbrush was named and linked to a size and number to identify them. The natural hairs of some paintbrushes have been dyed. A paintbrush wall support is suggested after you use the product, in order to keep your workspace clean and organized.

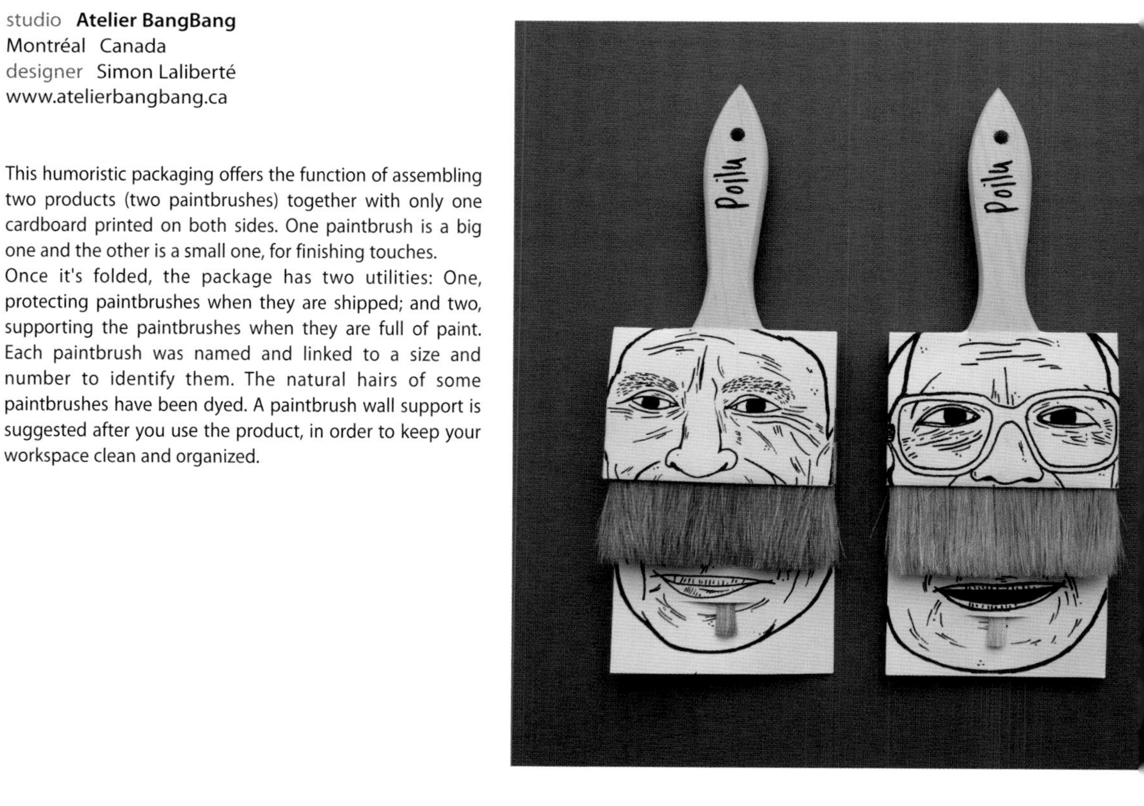

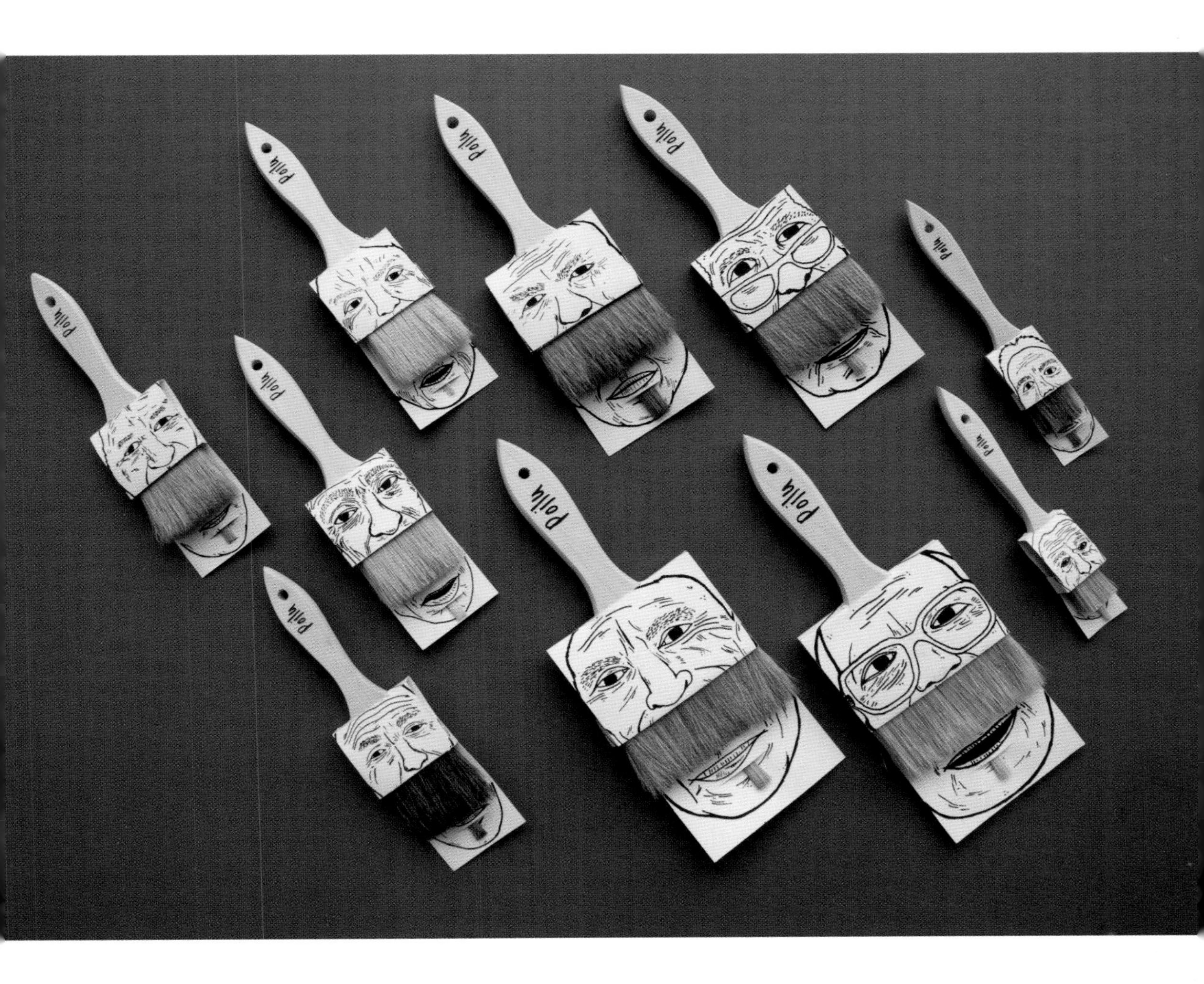

Ephrem
Poilu

studio **P&W Design Consultants**
Manhattan USA
designer Phil Curl
creative directors Simon Pemberton & Adrian Whitefoord
www.p-and-w.com

The three-strong range - Cookie Bites, Cocoa Sharks and Apple & Cinnamon Smiles - is designed to achieve cut-through in a category replete with bright colors, artificial flavors and cartoon characters.
We transformed the boxes into individual characters, each one about to guzzle back a bowl of their favorite cereal. Their bold, graphic style, is unique to the market and stands apart from the more visually cluttered branded competitors. To increase engagement, repeat purchase and imbue the packs with personality, we also created games, such as mazes and fun word-finders, to feature on the back of the pack.

Mother's little helpers

agency **Kolle Rebbe / KOREFE**
Hamburg Germany
creative director Antje Hedde
designer / idea Christine Knies
illustrator MASA
copywriter Gereon Klug
www.korefe.de

Housewives in the 1950s were well aware of "Mother's Little Helper" – relief for everyday tiredness, lethargy and stress – thought to have dubious medical effects. For its 'relaunch', "Mother's Little Helper" was reinvented as a chewing gum made solely of natural ingredients and packaged in a handbag-friendly tin. The packaging design complements the clever copy, both of which play on the individual gum flavor's promise. Every one of the four tins was lovingly illustrated and the individual pseudo illness is shown using psychedelic-like retro motifs and exaggerated symbols of femininity. The colors are in line with the original tranquilizer.

Mother's
LITTLE HELPERS
BALMY BASIL
CHEWING GUM
NATURAL
DOUBLE FLAVOURED
BASIL & LIME
BALMY BASIL
CHEWING GUM
* * *
THE BALMY POWER OF
BASIL MARRIED WITH LIME.
ILLUSTRATED WITH A LIGHT HAND – AND
FLYING SYMBOLS SUCH AS BUTTERFLIES,
BIRDS AND SMALL UMBRELLAS.
LIFE SUDDENLY BECOMES LIGHTER AND
THE PRESSURE SLIPS AWAY.
Mother's
LITTLE HELPERS
CHEERY CHOCOLATE
CHEWING GUM
NATURAL
DOUBLE FLAVOURED
CHOCOLATE & MINT
CHEERY CHOCOLATE
CHEWING GUM
* * *
THE EUPHORIC CHARACTER OF
CHOCOLATE REFINED BY FRESH MINT.
PLEASURE GIVES THE MOOD A BOOST:
A SENSUAL SEA OF CHOCOLATE
WAVES ROLLS IN.
LIFE CAN BE LIVED
AGAIN.

Mother's
LITTLE HELPERS
LAZY LAVENDER
CHEWING GUM
NATURAL
DOUBLE FLAVOURED
LAVENDER & WHITE PEACH
LAZY LAVENDER
CHEWING GUM
THE SOOTHING POWER OF
LAVENDER WITH WHITE PEACH.
CALMED, THE WONDER WOMAN CAN
PUT HER FEET UP AND RELAX.
TIME IS COMPLETELY FORGOTTEN. A MILD
COLOUR OF STIMULATION.
Mother's
LITTLE HELPERS
QUIRKY QUINCE
CHEWING GUM
NATURAL
DOUBLE FLAVOURED
QUINCE & SAFFRON
QUIRKY QUINCE
CHEWING GUM
THE REGENERATING POWER OF
QUINCE WITH GENUINE SAFFRON
WHAT A BOOST OF ENERGY! IT SEEMS AS IF
EVERYTHING IS POSSIBLE AND DOABLE.
EVEN WITH A SLIGHTLY EXAGGE-
RATED CRAZINESS.

agency **Gworkshop Design**
Quito Ecuador
designer José Luis García Eguiguren
www.gworkshopdesign.com

This particular toy is made with 100% biodegradable materials. The idea was conceived by the development of a character and by giving a visual differentiation on the color and expression.
Each particular character was given an adjective in order to convey a different product. This powerful and colorful pack attracts the attention of every kid in the store.

MR.
GOO
BLAZING
RED
NET WT: 120g
100% BIODEGRADABLE
MR.
GOO
SUPREME
GREEN
NET WT: 120g
100% BIODEGRADABLE
MR.
GOO
RELAXING
BLACK
NET WT: 120g
100% BIODEGRADABLE
MR.
GOO
CHILLING
BLUE
NET WT: 120g
100% BIODEGRADABLE
MR.
GOO
PUNCHED-OUT
PURPLE
NET WT: 120g
100% BIODEGRADABLE
MR.
GOO
EXTREME
ORANGE
NET WT: 120g
100% BIODEGRADABLE

MR.
GOO
EXTREME
ORANGE
NET WT: 120g
100% BIODEGRADABLE
MR.
GOO
EXTREME
ORANGE
NET WT: 120g
100% BIODEGRADABLE

Kefalonia Fisheries

agency **Mousegraphics**
Athens Greece
art director Greg Tsaknakis
food styling Tina Webb
illustrator Ioanna Papaioannou
photographer Dimitris Poupalos
www.mousegraphics.gr

We want to offer our clients our basic products - Kefalonia sea bass and sea bream - fresh, cleaned and ready to cook. The target audience: a wide but eclectic clientele. The design: we actually designed the consumer's next move, as far as the product is concerned. A fresh, cleaned and ready to be consumed fish does not require much except the simple act of cooking it with fine herbs that will bring forward its basic nutritious qualities. A differentiating band, on an otherwise transparent packaging, offers much more than the image of a serving suggestion. It works like an x-ray image of a pure product of Greek nature, as well as a preview of the particular culinary experience - the moment just before eating, when a fish is opened and all the fine tastes are ready to be liberated.

Ενέργεια
142kcal
596kJ
7%*
KEFALONIA
fisheries
organic
sea bream*
*gutted and scaled
ready to cook

KEFALONIA
fisheries
organic
sea bass*
*gutted and scaled
ready to cook

Fruta del Diablo

agency **Moxie Sozo**
Boulder, CO USA
designer Nate Dyer
www.moxiesozo.com

There are a wide variety of salsas in the marketplace, with offerings from small start-ups and international corporations vying for consumer dollars.
Moxie Sozo wanted to create salsa packaging for Fruta Del Diablo that would distinguish it from everything else on the shelf and establish credibility for an unknown brand. By using hand-drawn illustrations inspired by the woodcuts of Mexican artist Jose Guadalupe Posada, we were able to lend authenticity to the salsa while reinforcing the product's heritage in traditional Mexican cuisine.

UTA DEL DIABLO
SALSA

FRUTA DEL DIABLO
SALSA

SANTA FE. COSECHADAS CON
ABLO
SALSA
M
MADE IN SANTA FE

SALSA ORGÁNICA FRESCA HECHA EN SANTA FE. COSECHADAS
FRUTA DEL DIABLO
FRESH
SALSA
NET WEIGHT 16 OUNCES {454 g}
HOT
MADE IN SANTA FE

Bamboo Cherry et Coconut

agency **lg2boutique**
Montréal Canada
creative director Claude Auchu
graphic designer Anne-Marie Clermont
account services Catherine Lanctôt, Johanna Lessard
photography Mathieu Lévesque
www.lg2boutique.com

Although the colorful, happy image of Fruits and Passion is still present, the CHERRY, BAMBOO and COCONUT packaging strives to convey a more minimalist approach with a generous use of white space. This look is turning heads in-stores, where the art direction of the photography captures all the originality of this elegant and sophisticated universe.

BAMBOO
Fruits&passion
BAMBOO
Fruits&passion
BAMBOO
Fruits&passion

COCONUT
COCONUT
Fruits&passion
Fruits&passion
BAMBOO
Fruits&passion
BAMBOO
Fruits&passion
BAMBOO
Fruits&passion
CHERRY
Fruits&passion

agency **Kolle Rebbe / KOREFE**
Hamburg Germany
creative director Katrin Oeding
designer Christian Doering
copywriter Daniel Hoffmann
www.korefe.de

The Hamburg-based production agency Romey von Malottky brings color into the daily routine at the office. Which is why they sent out a set of teas as a Christmas mail-out in the colors of the 4C printing process. Packed exquisitely in small paint-pot tea tins – colors to smell and taste.

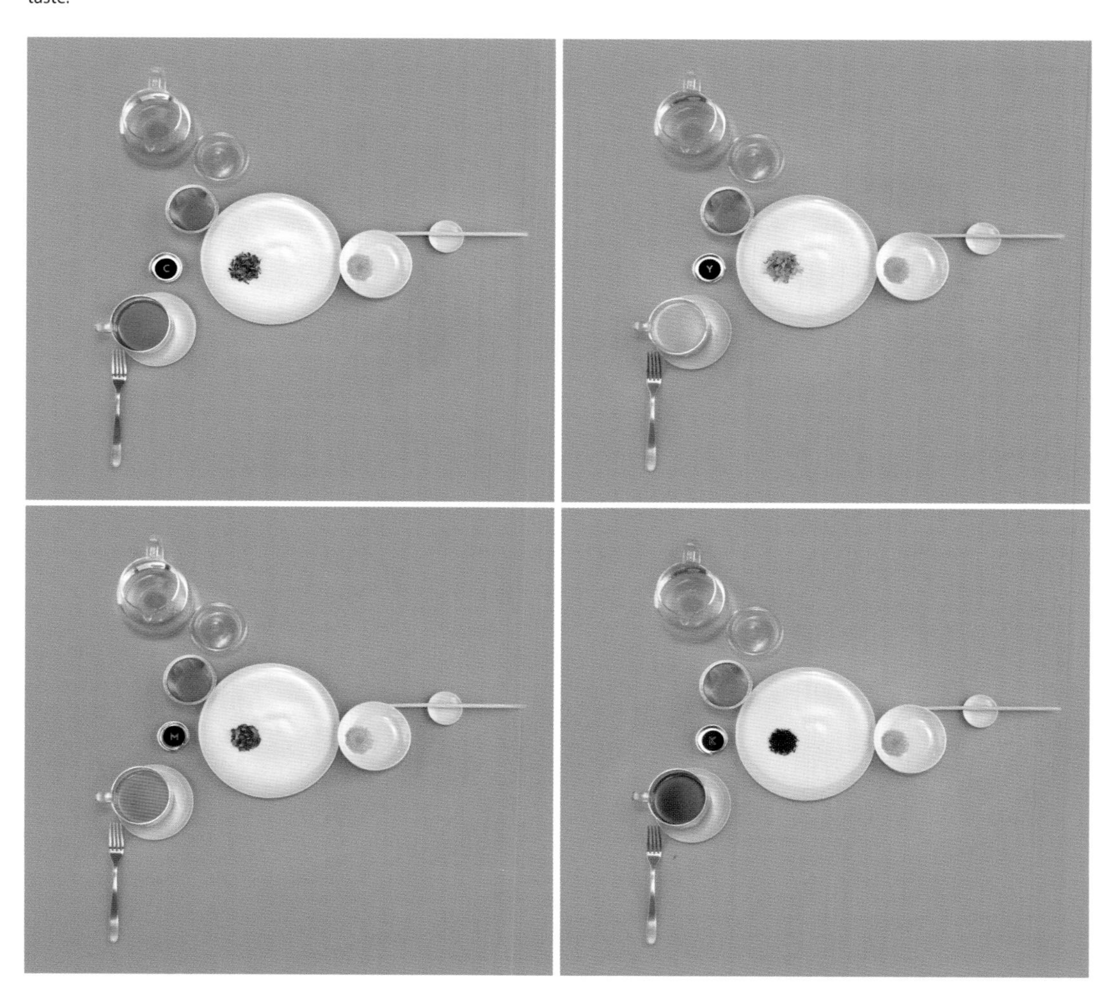

CMYK
TEE
C
Grüntee · Malve · Kornblume

CMYK
TEE
M
Rooibos · Himbeer · Pfeffer

CMYK
TEE
Y
Apfel · Ingwer · Süßholz

CMYK
TEE
K
Ceylon · Assam · Java Blatt

C

M

Y

K

agency **Cocoa Branding**
Jalisco México
designer Rodrigo Suárez Araiza
www.cocoabranding.com

Twïns is a conceptual, mid-range brand of men's underwear aimed at the18-25 market. This packaging series is designed in a fun, slightly-mischievous, retro-pop style to engage the target demographic and present a memorable solution. Cocoa Branding created a clever theme centered around 'the chick magnet' and executed it with tongue-in-cheek style, creating packaging that's whimsical and original. Playing with the double entendre of common sayings, this line of packaging tells a different story on each box. On the front, the box shows a clear window so the product can be seen and touched, along with an illustration reminiscent of school-book anatomy illustrations.

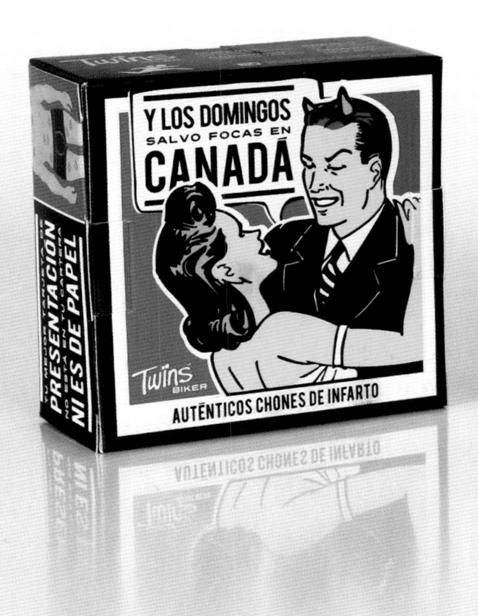
Y LOS DOMINGOS
SALVO FOCAS EN
CANADÁ
Twins
BIKER
AUTÉNTICOS CHONES DE INFARTO
PRESENTACIÓN
NI ES DE PAPEL

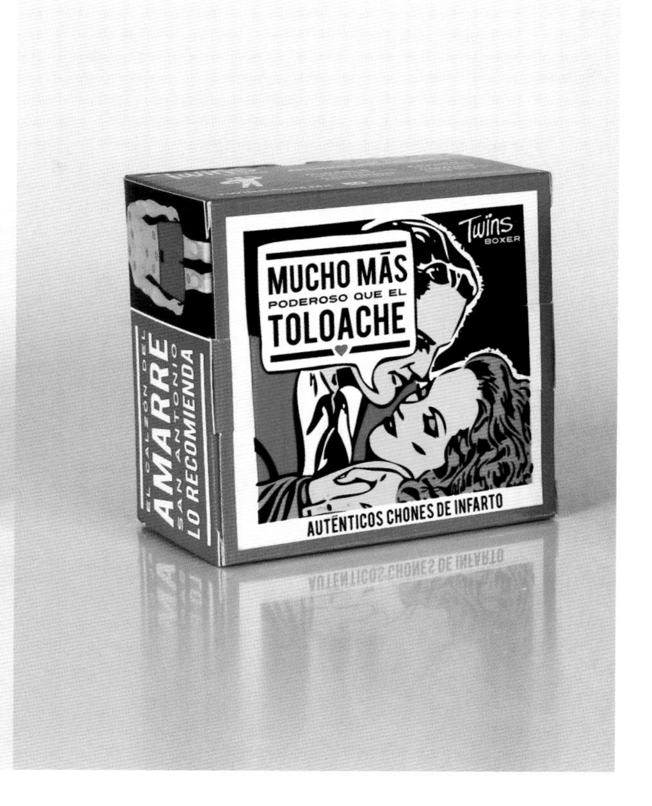
Twins
BOXER
MUCHO MÁS
PODEROSO QUE EL
TOLOACHE
AUTÉNTICOS CHONES DE INFARTO
AMARRE
LO RECOMIENDA

PORQUE LO TUYO
SIEMPRE HA SIDO
LA QUÍMICA
Twins
BIKER
TUS CLASES
DE ANATOMÍA

Twins
BIKER
TWINS.COM.MX
AUTÉNTICOS CHONES DE INFARTO

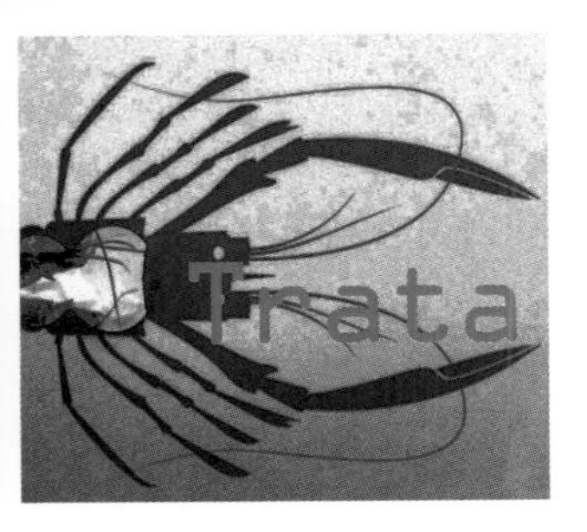

Trata OnIce

agency **Beetroot Design Group**
Thessaloniki Greece
designers Vagelis Liakos, Alexis Nikou & Yiannis Charalambopoulos
www.beetroot.gr

The tail of the fish was one of the main visual elements that was used on this frozen seafood packaging series. It was used on both the design of the logo and the structure of the packages. It is a very strong element of the brand and makes the products recognizable from any angle. The black background, the subtle typography and the seafood illustrations were used in order to underline the quality of the product and give it a delicatessen feeling. You can actually see the product through the holes on the illustrations and they highlight the part of the seafood that is included inside the package.

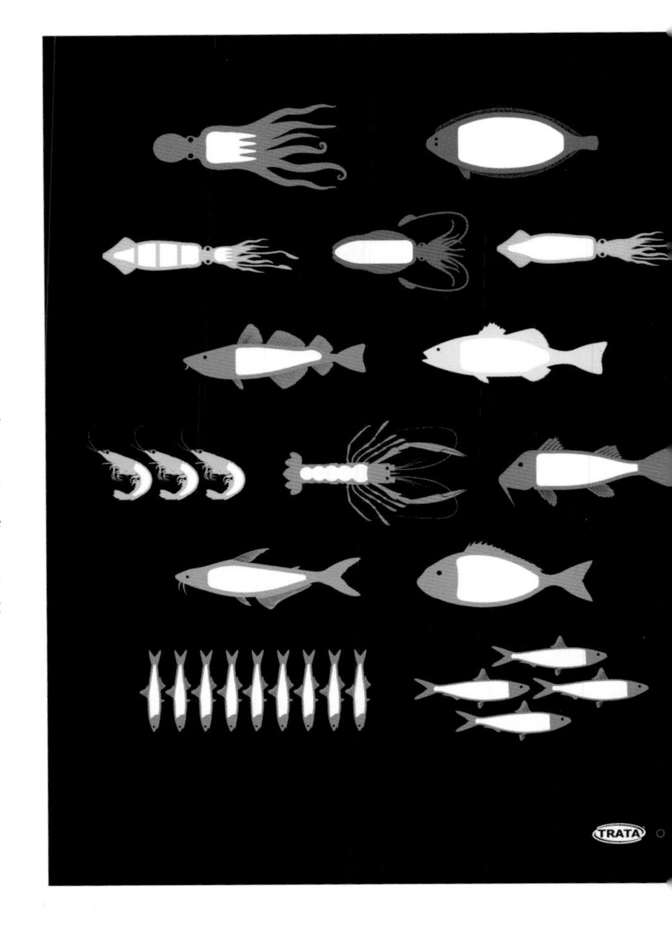

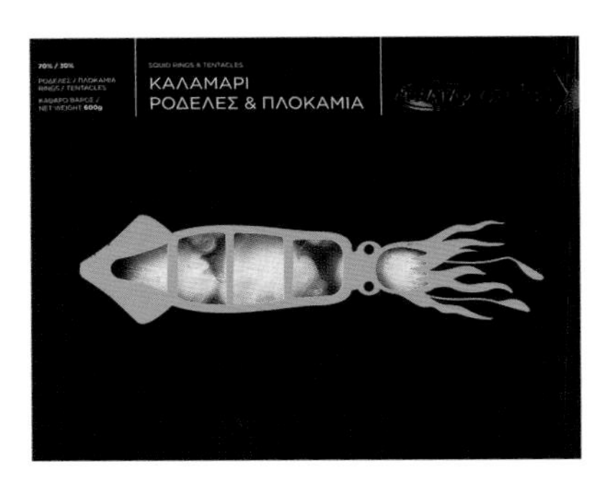

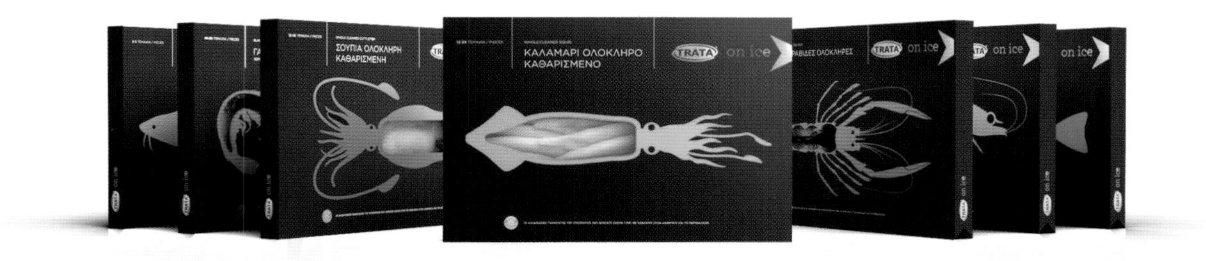
ΣΟΥΠΙΑ ΟΛΟΚΛΗΡΗ
ΚΑΘΑΡΙΣΜΕΝΗ
ΚΑΛΑΜΑΡΙ ΟΛΟΚΛΗΡΟ
ΚΑΘΑΡΙΣΜΕΝΟ
TRATA
on ice

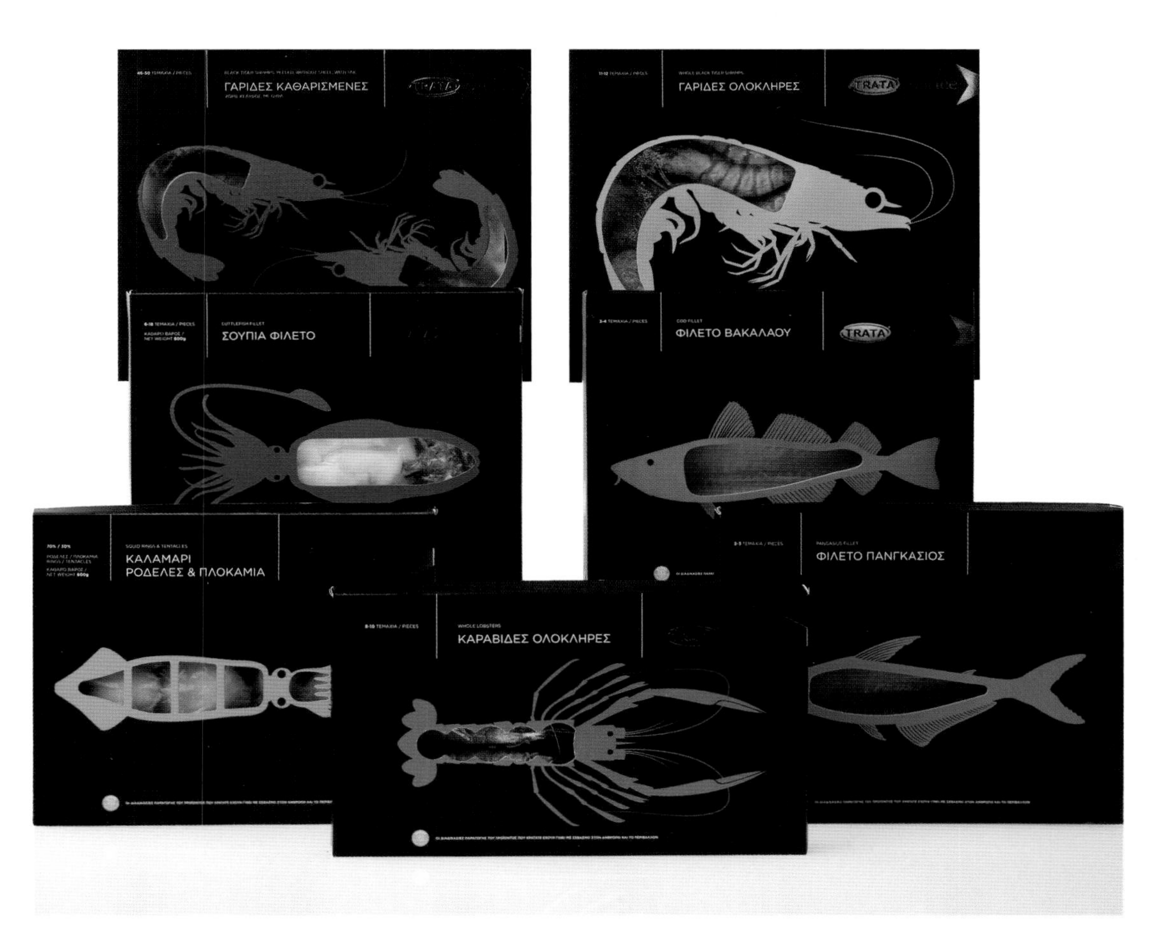
ΓΑΡΙΔΕΣ ΚΑΘΑΡΙΣΜΕΝΕΣ
TRATA
ΓΑΡΙΔΕΣ ΟΛΟΚΛΗΡΕΣ
TRATA
ΣΟΥΠΙΑ ΦΙΛΕΤΟ
ΦΙΛΕΤΟ ΒΑΚΑΛΑΟΥ
TRATA
ΚΑΛΑΜΑΡΙ
ΡΟΔΕΛΕΣ & ΠΛΟΚΑΜΙΑ
ΦΙΛΕΤΟ ΠΑΝΓΚΑΣΙΟΣ
ΚΑΡΑΒΙΔΕΣ ΟΛΟΚΛΗΡΕΣ

Quick Fruit

studio **Marcel Buerkle**
Johannesburg South Africa
designers Marcel Buerkle
www.behance.net/marcel_b

The idea behind Quick Fruit packaging is a fruit sliced in half, showing the core of the fruit as the lid of the product. A clean, simple logo with the letter "Q" depicting a cup with a spoon appears on the lid and side.

Hill Station

agency **Williams Murray Hamm**
London United Kingdom
creative director Garrick Hamm
designer Fiona Curran
www.williamsmurrayhamm.com

The Hill Station brand was based on a charming true story of British exploits. Founders Charles and Gina Hall, an American entrepreneurial couple, had travelled extensively and sampled fruits and produce from all around the world. Their original brand concept was inspired by the historical notion of Britain's travelling in the tropics (Hill Stations were a retreat dating from colonial times where people would go to escape the oppressive heat). The redesign took this as its reference point and used characteristic hand-painted signage, set against evocative backdrops of tropical skies, to create a memory of their travels.

Seriously Good

agency **Williams Murray Hamm**
London United Kingdom
creative director Garrick Hamm
designer/typography Emma Slater
photography Ray Burmiston
account manager Kate Spence
www.williamsmurrayhamm.com

The brief was to create some noise in the ambient cooking sauce category by creating, designing and naming a new sauce brand, devised by Gordon Ramsay to benefit Comic Relief.
This will be bought by busy adults looking for a quick and easy way to create a meal and feel good about supporting the Comic Relief charity. They aspire to do real cooking, with real ingredients but don't always have the time.
This is a dull category. Convention ridden - lots of red and green jars and the odd splash of silver to signal premium-ness. The design uses the mischievous spirit of Comic Relief to provide an antidote to both a dull category and celebrity chef overload. Gordon's nose is covered in red sauce, which is in contrast with his Michelin starred celebrity chef image. This 'sauce accident' also resembles Comic Relief's logo of a clown's red nose. The name, Seriously Good, builds on Gordon's direct use of language.

Emmi Swiss Yoghurt

agency **Studio h**
London United Kingdom
design & illustration Rob Hall
www.studioh.co.uk

New product development for a UK launch of Emmi Swiss yogurt uses the fresh, clean, minimal design attributes that Switzerland is famous for to create a powerful brand identity that comes to life when inhabited by the cow.

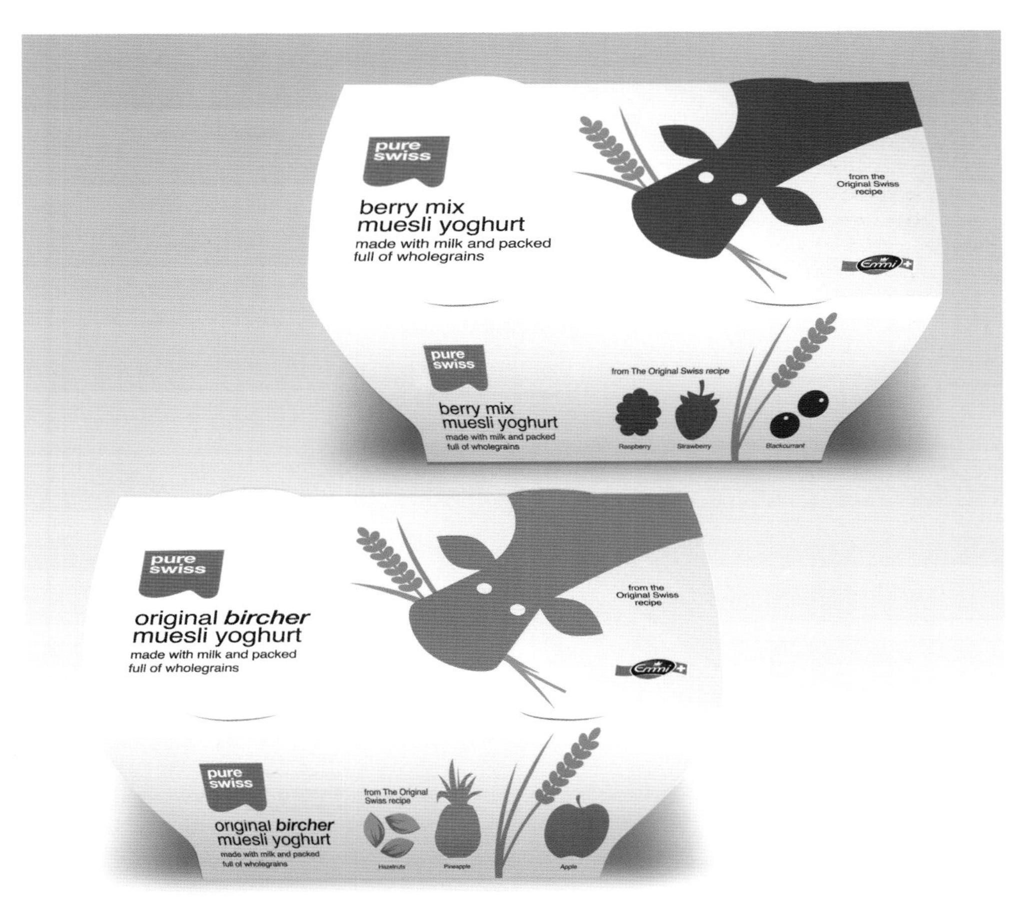

Chilly Moo

Dublin Ireland
designer Steve Simpson
www.stevesimpson.com

Chilly Moo is a frozen yogurt start-up company based in Dublin, Ireland. They wanted a design that would appeal to kids and at the same time say "healthy" to parents.
The initial range consists of 3 flavors featuring sweater wearing cows playing.

CHILLYMOO
FROZEN YOGURT
Banana & Strawberry

CHILLYMOO
FROZEN YOGURT
Mixed Berry

CHILLYMOO
FROZEN YOGURT
Strawberry

studio **Istratova Alexandra**
Moscow Russia
designer Istratova Alexandra
www.behance.net/Sasha-Tyla

This packing is for a puree of vegetables and fruits. Every 'kiss' has its own unique taste and feeling... you can find what you like! The packs are an unusual mix of natural fruits and vegetables and are great for consuming on the go.

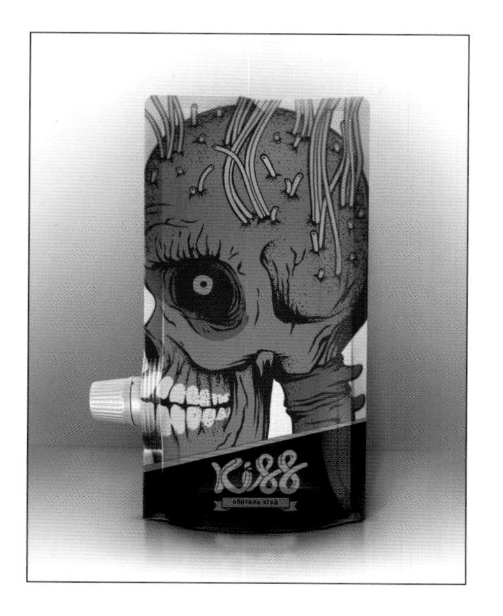

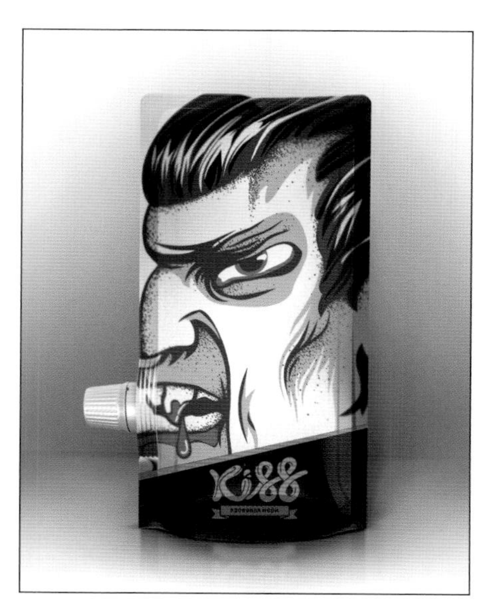

Kiss
дикие овощи

Kiss

Kiss

Triptea

studio **Andrew Gorkovenko**
Moscow Russia
designer Andrew Gorkovenko
www.gorkovenko.ru

At the heart of brand communication is the idea of traveling to exotic countries around the world, which leads us TripTea and opens up a new world of flavor with each new package.
Since the brand's communication and packaging were designed simultaneously, the semantic component was reflected in all cells at once. Naming the brand was based on two strong associations: first, TripTea is a tea travel, and secondly, TripTea tunes in the familiar and figurative word "triptych" - a three-part painting that completely reveals one key message in different subjects. In developing the package, this principle is also embodied. We wanted to show the beauty, depth and fullness of the tea exclusively. Therefore, the packaging was decorated with landscapes of countries where it was assembled and produced. All landscapes are handmade directly from the tea variety in the package. This conveys an exotic image of the country along with the richness of flavors and nuances of the product itself.

BLACK
triptea
Kenya
GREEN
triptea
BLACK
triptea
Ceylon

BLACK
triptea
Kenya
Black tea
Tea production in Kenya
100 g
Ceylon tea is specially selected for creative tea parties
triptea

Soup Delhaize

agency **Lavernia & Cienfuegos**
Valencia Spain
www.lavernia-cienfuegos.com

The brief was to bring to life the principle ingredient, preferably through the use of photo-realistic images, with something that adds a touch of good humor before serving.
The romantic image of the waiter's hands gives positive connotations of good service and quality. The steady black and white photography helps to visually unify the whole range, which is complimented by the use of simple bold typography to balance the design. The size of the ingredients, in color, have been exaggerated in relation to the plate to emphasise the high natural vegetable content of the soups in comparison to its competitor's. This play-on-size, coupled with the waiter concept, adds a touch of humor. Bon appétit!

DELHAIZE
SOUP

DELHAIZE
SOUP

DELHAIZE
SOUP

DELHAIZE
SOUP
Vissoep

DELHAIZE
SOUP

DELHAIZE
SOUP

agency **Blue Marlin Brand Design**
London UK
www.bluemarlinbd.com

Green Saffron is the passion project of entrepreneur Arun Kapil, whose ambition is to bring the freshest, highest-grade whole spices straight from the farms of India to the consumer's kitchen. Initially sold at farmers' markets, the brand's reputation grew rapidly. Blue Marlin was engaged to create a uniquely personal brand identity, develop strategic and visual positioning and design packaging for the spice specialist, to enable the spice specialist to join the mainstream at shelf.

Mad Ink

studio **The British Higher School of Art and Design**
Moscow Russia
con?ept & design Shashkina Ivanna
ivanna.shashkina@gmail.com

This is a line of hair care products for young people. It includes shampoo, conditioner and styling products. The main component is washable hair dye. You can change your image of a single evening. Mad Ink expresses rage and energy, as well as disobedience to standards.

0 1341 1573556 2

KEFALONIA
organic
sea bream*
*gutted and sealed
ready to cook